Total Quality Management

Pacific Rim Edition

For Susan and Harpreet

Total Quality Management

Pacific Rim Edition

Text with cases

John S Oakland

PhD, CChem, MRSC, FIQA, FSS, MASQC

Exxon Chemical Professor of
Total Quality Management,
European Centre for TQM,
Management Centre, University of Bradford, UK

Amrik S Sohal

PhD, MBA, BEng (Hons), CEng, MIEE, MIQA

Professor and Associate Dean (Graduate Teaching)
Faculty of Business and Economics, Monash University,
Melbourne, Australia

Published 1996 by
Butterworth-Heinemann Australia
an imprint of Reed Reference Australia, part of
Reed Educational & Professional Publishing
18–22 Salmon Street, Port Melbourne
Victoria 3207
Australia

A division of Reed International Books Pty Ltd ACN 001002357

A Reed Elsevier company

© text John S Oakland, 1995
© cases Amrik S Sohal, 1996

This Pacific Rim edition is based on John S Oakland's *Total Quality Management: Text with cases*
(Butterworth-Heinemann Oxford, 1995).

National Library of Australia Cataloguing-in-Publication data

Sohal, Amrik.
    Total quality management.

    Asia/Pacific ed.
    Includes index.
    ISBN 0 7506 8925 0.

    1. Total quality management. 2. Production management – Quality control.
    3. Quality assurance. I. Oakland, John S. Total quality management. II. Title.

658.5

Typeset in Australia by J&M Typesetting
Cover design by Cate Mills
Printed in Australia by McPherson's Printing Group, Maryborough

Contents

Preface

Continuous cost reduction, productivity and quality improvement have proved essential for organisations to stay in operation. We cannot avoid seeing how quality has developed into the most important competitive weapon, and many organisations have realised that TQM is *the* way of managing for the future. TQM is far wider in its application than assuring product or service quality – it is a way of managing business processes to ensure complete 'customer' satisfaction at every stage, internally and externally.

This book is about how to manage in a total quality way. It is based on the very successful book *Total Quality Management* by John Oakland (published by Butterworth-Heinemann, second edition, 1993). The book is structure around five parts of a model for TQM. The core of the model is the *customer–supplier* interfaces, both externally and internally, and the fact that at each interface there lies a number of *processes*. This sensitive core must be surrounded by commitment to quality, meeting the customer requirements, communication of the quality message, and recognition of the need to change the culture of most organisations to create total quality. These are the soft FOUNDATIONS, to which must be added the SYSTEMS, the TOOLS, and the TEAMS – the hard management necessities.

Under these headings the essential steps for the successful IMPLEMENTATION of TQM are set out in what is a meaningful and practical way. The book should guide the reader through the language of TQM and sets down a clear way to proceed for organisations.

Many of the 'gurus' appear to present different theories of quality management. In reality they are talking the same 'language' but they use different dialects; the basic principles of defining quality and taking it into account throughout all the activities of the 'business' are common. Quality has to be managed – it does not just happen. Understanding and commitment by senior management, effective leadership and teamwork are fundamental parts of the recipe for success. This success in implementation is demonstrated through 11 case studies from the Pacific Rim region. Four case studies are from Australia, two each from Hong Kong and New Zealand and one each from Fiji, Malaysia and Singapore.

The book should meet the requirements of the increasing number of students who need to understand the part TQM may play in their courses on science, engineering, or management. We hope that those engaged in the pursuit of professional qualifications in the management of quality assurance, such as membership of the Institute of Quality Assurance, the American Society for Quality Control, the Australian Quality Council and the Australian Organisation for Quality, will make this book an essential part of their library.

We would like to thank our colleagues at the European Centre for TQM, University of Bradford, UK, and the Quality Management Research Unit, Monash University, Melbourne, Australia, for the sharing of ideas and help in their development. The book is the result of many years of collaboration in assisting organisations to introduce good methods of management and embrace the concepts of total quality. We are most grateful to MCB University Press for permission to publish some of the case studies and to their authors.

Amrik Sohal
May 1996

Part One

The Foundations – A Model for TQM

Good order is the foundation of all good things.

Edmund Burke, 1791

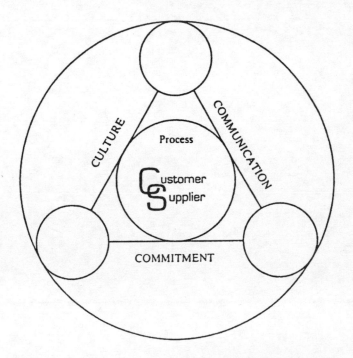

1
Understanding quality

1.1 Quality and competitiveness

There was a company outside Japan that tried to manufacture motor cars. Its name does not matter – it has changed its name now anyway – let's call it B Motors. The company did a deal with a Japanese car manufacturer that allowed the Japanese-named cars, let's call them HO, to be manufactured by B Motors. What happened in the boardroom of HO motor cars when that deal was announced can only be described as mass hara-kiri. Clearly, rather a lot of depression had settled over those particular Japanese gentlemen. Why should they behave in this way? What was the difference between the two companies? In a word, image or reputation. Reputation for what? Reputation for quality, reliability, price, and delivery – all the things we *compete* on.

Whatever type of organisation you work in – a hospital, a university, a bank, an insurance company, local government, an airline, a factory – competition is rife: competition for customers, for students, for patients, for resources, for funds. There are very few people around in most types of organisation who remain to be convinced that quality is the most important of the competitive weapons. If you doubt that, just look at the way some organisations, even whole industries in certain countries, have used quality to take the heads off their competitors. And they are not only Japanese companies. Australian, British, American, European, South Korean, Singaporean, Taiwanese organisations, and organisations from other countries, have used quality strategically to win customers, steal business resources or funding, and be competitive. Moreover, attention to quality improves performance in reliability, delivery, and price.

The reputation of Japanese companies once was anything but good. Not too long ago they were most famous for 'cheap oriental trash'. They have clearly *learned* something. This has not as much to do with differences in national cultures as many people think it has. The Japanese culture, which is much older than most western cultures has not changed significantly in 40–50 years.

One of the lessons many Japanese companies learned after the Second World War was to manage quality, and the other things on which we compete. They learned it from a handful of Americans – people like Joseph M Juran and W Edwards Deming, who have since reached fame as 'gurus' of quality management.

The company we called B Motors also learned a thing or two about quality and competition. It lost market share, but started to put things right by a better understanding of quality management and the needs of its customers. Unfortunately its previous reputation was so bad that it is taking it many years to change people's view. It may never do so. Moreover, the country in which it operates gained a poor reputation for shoddy goods and services, in contrast to the 'Japanese', who seem to take so many industries by storm. Even the trains run on time there!

For any organisation, there are several lessons to be learned about reputation from this story:

1 It is built upon the competitive elements of quality, reliability, delivery, and price, of which quality has become strategically the most important.
2 Once an organisation acquires a poor reputation for quality, it takes a very long time to change it.
3 Reputations, good or bad, can quickly become national reputations.
4 The management of the competitive weapons, such as quality, can be learned like any other skill, and used to turn round a poor reputation, in time.

Before anyone will buy the idea that quality is an important consideration, they would have to know what was meant by it.

What is quality?

'Is this a quality watch?' Pointing to your wrist, you ask this question of a class of students – undergraduates, postgraduates, experienced managers – it matters not who. The answers you receive will vary:

• 'No, it's made in Japan.'
• 'No, it's cheap.'
• 'No, the face is scratched.'
• 'How reliable is it?'
• 'I wouldn't wear it.'

Very rarely you will receive an answer which states that the quality of the watch depends on what the wearer requires from a watch – perhaps a piece of jewellery to give an impression of wealth; a timepiece that gives the required data, including the date, in digital form; or one with the ability to perform at 50 metres under the sea? Clearly these requirements determine the quality.

Quality is often used to signify 'excellence' of a product or service – people talk about 'Rolls-Royce quality' and 'top quality'. In some engineering companies the word may be used to indicate that a piece of metal conforms to certain physical dimensional characteristics often set down in the form of a particularly 'tight' specification. In a hospital it might be used to indicate some sort of 'professionalism'. If we are to define quality in a way that is useful in its *management*, then we must recognise the need to include in the assessment of quality the true requirements of the 'customer' – the needs and expectations.

Quality then is simply *meeting the customer requirements*, and this has been expressed in many ways by other authors:

• 'Fitness for purpose or use' – Juran.
• 'The totality of features and characteristics of a product or service that bear on its

ability to satisfy stated or implied needs' – (ISO 8402, 1986) *Quality Vocabulary*: Part 1, *International Terms*.
- 'Quality should be aimed at the needs of the consumer, present and future' – Deming.
- 'The total composite product and service characteristics of marketing, engineering, manufacture and maintenance through which the product and service in use will meet the expectation by the customer' – Feigenbaum.
- 'Conformance to requirements' – Crosby.

Another word that we should define properly is *reliability*. 'Why do you buy a Honda car?' 'Quality and reliability' comes back the answer. The two are used synonymously, often in a totally confused way. Clearly, part of the acceptability of a product or service will depend on its ability to function satisfactorily *over a period of time*, and it is this aspect of performance that is given the name *reliability*. It is the ability of the product or service to *continue* to meet the customer requirements. Reliability ranks with quality in importance, since it is a key factor in many purchasing decisions where alternatives are being considered. Many of the general management issues related to achieving product or service quality are also applicable to reliability.

It is important to realise that the 'meeting the customer requirements' definition of quality is not restrictive to the functional characteristics of products or services. Anyone with children knows that the quality of some of the products they purchase is more associated with *satisfaction in ownership* than some functional property. This is also true of many items, from antiques to certain items of clothing. The requirements for status symbols account for the sale of some executive cars, certain bank accounts and charge cards, and even hospital beds! The requirements are of paramount importance in the assessment of the quality of any product or service.

By *consistently* meeting customer requirements, we can move to a different plane of satisfaction – *delighting the customer*. There is no doubt that many organisations have so well ordered their capability to meet their customers' requirements, time and time again, that this has created a reputation for 'excellence'.

1.2 Understanding and building the quality chains

The ability to meet the customer requirements is vital, not only between two separate organisations, but within the same organisation.

When the air hostess pulled back the curtain across the aisle and set off with a trolley full of breakfasts to feed the early morning travellers on the short domestic flight into an international airport, she was not thinking of quality problems. Having stopped at the row of seats marked 1ABC, she passed the first tray onto the lap of the man sitting by the window. By the time the second tray had reached the lady beside him, the first tray was on its way back to the air hostess with a complaint that the bread roll and jam were missing. She calmly replaced it in her trolley and reached for another – which also had no roll and jam.

The calm exterior of the girl began to evaporate as she discovered two more trays without a complete breakfast. Then she found a good one and, thankfully, passed it over. This search for complete breakfast trays continued down the aeroplane, causing inevitable delays, so much so that several passengers did not receive their breakfasts

until the plane had begun its descent. At the rear of the plane could be heard the mutter-
ings of discontent. 'Aren't they slow with breakfast this morning?' 'What is she doing
with those trays?' 'We will have indigestion by the time we've landed.'

The problem was perceived by many to be one of delivery or service. They could
smell food but they weren't getting any of it, and they were getting really wound up!
The air hostess, who had suffered the embarrassment of being the purveyor of defective
product and service, was quite wound up and flushed herself, as she returned to the
curtain and almost ripped it from the hooks in her haste to hide. She was heard to say
through clenched teeth, 'What a bloody mess!'

A problem of quality? Yes, of course, requirements not being met, but where? The
passengers or customers suffered from it on the aircraft, but down in the bowels of the
organisation there was a little man whose job it was to assemble the breakfast trays. On
this day the system had broken down – perhaps he ran out of bread rolls, perhaps he
was called away to refuel the aircraft (it was a small airport!), perhaps he didn't know
or understand, perhaps he didn't care.

Three hundred Km away in a chemical factory ... 'What the hell is Quality Control
doing? We've just sent 15,000 litres of lawn weedkiller to CIC and there it is back at
our gate – they've returned it as out of spec.' This was followed by an avalanche of
verbal abuse, which will not be repeated here, but poured all over the shrinking Quality
Control Manager as he backed through his office door, followed by a red faced
Technical Director advancing menacingly from behind the bottles of sulphuric acid
racked across the adjoining laboratory.

'Yes, what is QC doing?' thought the Production Manager, who was behind a door
two offices along the corridor, but could hear the torrent of language now being used to
beat the QC man into an admission of guilt. He knew the poor devil couldn't possibly
do anything about the rubbish that had been produced except test it, but why should he
volunteer for the unpleasant and embarrassing ritual now being experienced by his
colleague – for the second time this month. No wonder the QC manager had been
studying the employment pages of the national newspapers – what a job!

Do you recognise these two situations? Do they not happen every day of the week –
possibly every minute somewhere in manufacturing or the service industries? Is it any
different in banking, insurance, the health service? The inquisition of checkers and
testers is the last bastion of desperate systems trying in vain to catch mistakes, stop
defectives, hold lousy materials, before they reach the external customer – and woe
betide the idiot who lets them pass through!

Two everyday incidents, but why are events like these so common? The answer is the
acceptance of one thing – *failure*. Not doing it right the first time at every stage of the
process.

Why do we accept failure in the production of artefacts, the provision of a service, or
even the transfer of information? In many walks of life we do not accept it. We do not
say, 'Well, the nurse is bound to drop the odd baby in a thousand – it's just going to
happen'. We do not accept that!

In each department, each office, even each household, there are a series of suppliers
and customers. The typist is a supplier to her boss. Is she meeting his requirements?

Does he receive error-free typing set out as he wants it, when he wants it? If so, then
we have a quality typing service. Does the air hostess receive from her supplier in the
airline the correct food trays in the right quantity?

Outside organisation

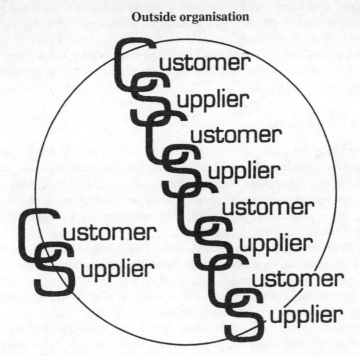

Figure 1.1 *The quality chains*

Throughout and beyond all organisations, whether they be manufacturing concerns, banks, retail stores, universities, hospitals or hotels, there is a series of *quality chains* of customer and suppliers (Figure 1.1) that may be broken at any point by one person or one piece of equipment not meeting the requirements of the customer, internal or external. The interesting point is that this failure usually finds its way to the interface between the organisation and its outside customers, and the people who operate at that interface – like the air hostess – usually experience the ramifications. The concept of internal and external customers/suppliers forms the *core* of total quality.

A great deal is written and spoken about employee motivation as a separate issue. In fact the key to motivation *and* quality is for everyone in the organisation to have well-defined customers – an extension of the word beyond the outsider that actually purchases or uses the ultimate product or service to anyone to whom an individual gives a part, a service, information – in other words the results of his or her work.

Quality has to be managed – it will not just happen. Clearly it must involve everyone in the process and be applied throughout the organisation. Many people in the support functions of organisations never see, experience, or touch the products or services that their organisations buy or provide, but they do handle or produce things like purchase orders or invoices. If every fourth invoice carries at least one error, what image of quality is transmitted!

Failure to meet the requirements in any part of a quality chain has a way of multiplying, and failure in one part of the system creates problems elsewhere, leading to yet more failure, more problems and so on. The price of quality is the continual examination of the requirements and our ability to meet them. This alone will lead to a 'continuing improvement' philosophy. The benefits of making sure the requirements are met at every stage, every time, are truly enormous in terms of increased competitiveness and market share, reduced costs, improved productivity and delivery performance, and the elimination of waste. The Japanese have called this 'company-wide quality improvement' or CWQI.

Meeting the requirements

If quality is meeting the customer requirements, then this has wide implications. The requirements may include availability, delivery, reliability, maintainability and cost-effectiveness, among many other features. The first item on the list of things to do is find out what the requirements are. If we are dealing with a customer/supplier relationship crossing two organisations, then the supplier must establish a 'marketing' activity charged with this task.

The marketers must of course understand not only the needs of the customer but also the ability of their own organisation to meet them. If your customer places a requirement on you to run 1,500 metres in 4 minutes, then you should know whether or not you are able to meet this demand. You may never be able to achieve this target but you should be able to improve your running performance.

Within organisations, between internal customers and suppliers, the transfer of information regarding requirements is frequently poor to totally absent. How many executives really bother to find out what their customers' – their secretaries' – requirements are? Can their handwriting be read, do they leave clear instructions, do the secretaries always know where the boss is? Equally, do the secretaries establish what their bosses need – error-free typing, clear messages, a tidy office? Internal supplier/customer relationships are often the most difficult to manage in terms of establishing the requirements. To achieve quality throughout an organisation, each person in the quality chain must interrogate every interface as follows:

Customers
- Who are my immediate customers?
- What are their true requirements?
- How do or can I find out what the requirements are?
- How can I measure my ability to meet the requirements?
- Do I have the necessary capability to meet the requirements? (If not, then what must change to improve the capability?)
- Do I continually meet the requirements? (If not, then what prevents this from happening, when the capability exists?)
- How do I monitor changes in the requirements?

Suppliers
- Who are my immediate suppliers?
- What are my true requirements?
- How do I communicate my requirements?
- Do my suppliers have the capability to measure and meet the requirements?

- How do I inform them of changes in the requirements?

The measurement of capability is extremely important if the quality chains are to be formed within and without an organisation. Each person in the organisation must also realise that the supplier's needs and expectations must be respected if the requirements are to be fully satisfied.

To understand how quality may be built into a product or service, at any stage, it is necessary to examine the two distinct, but interrelated aspects of quality:

- Quality of design
- Quality of conformance to design.

Quality of design

We are all familiar with the old story of the tree swing (Figure 1.2), but in how many places in how many organisations is this chain of activities taking place? To discuss the quality of, say, a chair it is necessary to describe its purpose. What it is to be used for? If it is to be used for watching TV for 3 hours at a stretch, then the typical office chair will not meet this requirement. The difference between the quality of the TV chair and the office chair is not a function of how it was manufactured, but its *design*.

Quality of design is a measure of how well the product or service is designed to achieve the agreed requirements. The beautifully presented gourmet meal will not necessarily please the recipient if he or she is travelling on the highway and has stopped for a quick bite to eat. The most important feature of the design, with regard to achieving quality, is the specification. Specifications must also exist at the internal supplier/customer interfaces if one is to pursue company-wide quality. For example, the company lawyer asked to draw up a contract by the sales manager requires a specification as to its content:

1 Is it a sales, processing or consulting type of contract?
2 Who are the contracting parties?
3 In which countries are the parties located?
4 What are the products involved (if any)?
5 What is the volume?
6 What are the financial, eg price, escalation, aspects?

The financial controller must issue a specification of the information he or she needs, and when, to ensure that foreign exchange fluctuations do not cripple the company's finances. The business of sitting down and agreeing a specification at every interface will clarify the true requirements and capabilities. It is the vital first stage for a successful total-quality effort.

There must be a corporate understanding of the organisation's quality position in the market place. It is not sufficient that marketing specifies the product or service 'because that is what the customer wants'. There must be an agreement that the operating departments can achieve that requirement. Should they be incapable of doing so, then one of two things must happen: either the organisation finds a different position in the market place or substantially changes the operational facilities.

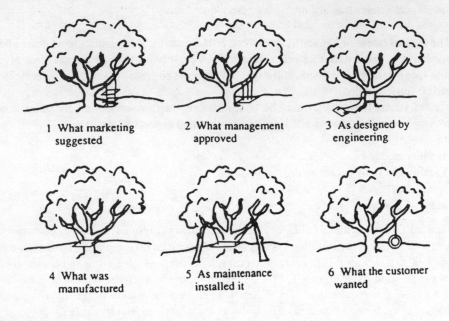

1 What marketing suggested

2 What management approved

3 As designed by engineering

4 What was manufactured

5 As maintenance installed it

6 What the customer wanted

Figure 1.2   *Quality of design*

Quality of conformance to design

This is the extent to which the product or service achieves the quality of design. What the customer actually receives should conform to the design, and operating costs are tied firmly to the level of conformance achieved. Quality cannot be inspected into products or services; the customer satisfaction must be designed into the whole system. The conformance check then makes sure that things go according to plan.

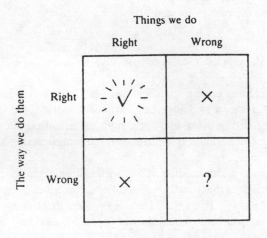

Figure 1.3   *How much time is spent on doing the right things right?*

A high level of inspection or checking at the end is often indicative of attempts to inspect in quality. This may well result in spiralling costs and decreasing viability. The area of conformance to design is concerned largely with the quality performance of the actual operations. It may be salutary for organisations to use the simple matrix of Figure 1.3 to assess how much time they spend doing the right things right. A lot of people, often through no fault of their own, spend a good proportion of the available time doing the right things wrong. There are people (and organisations) who spend time doing the wrong things very well, and even those who occupy themselves doing the wrong things wrong, which can be very confusing!

1.3 Managing processes

Every day two men who work in a certain factory scrutinise the results of the examination of the previous day's production, and begin the ritual battle over whether the material is suitable for despatch to the customer. One is called the Production Manager, the other the Quality Control Manager. They argue and debate the evidence before them, the rights and wrongs of the specification, and each tries to convince the other of the validity of his argument. Sometimes they nearly start fighting.

This ritual is associated with trying to answer the question, *'Have we done the job correctly?'*, correctly being a flexible word, depending on the interpretation given to the specification on that particular day. This is not quality *control*, it is *detection* – wasteful detection of bad product before it hits the customer. There is still a belief in some quarters that to achieve quality we must check, test, inspect or measure – the ritual pouring on of quality at the end of the process. This is nonsense, but it is frequently practised. In the office one finds staff checking other people's work before it goes out, validating computer input data, checking invoices, typing, etc. There is also quite a lot of looking for things, chasing why things are late, apologising to customers for lateness, and so on. Waste, waste, waste!

To get away from the natural tendency to rush into the detection mode, it is necessary to ask different questions in the first place. We should not ask whether the job has been done correctly, we should ask first *'Are we capable of doing the job correctly?'* This question has wide implications, and this book is devoted largely to the various activities necessary to ensure that the answer is yes. However, we should realise straight away that such an answer will only be obtained by means of satisfactory methods, materials, equipment, skills and instruction, and a satisfactory 'process'.

What is a process?

As we have seen, quality chains can be traced right through the business or service processes used by any organisation. A process is the transformation of a set of inputs, which can include actions, methods and operations, into outputs that satisfy customer needs and expectations, in the form of products, information, services or – generally – results. Everything we do is a process, so in each area or function of an organisation there will be many processes taking place. For example, a finance department may be engaged in budgeting processes, accounting processes, salary and wage processes, costing processes, etc. Each process in each department or area can be analysed by an examination of the inputs and outputs. This will determine some of the actions necessary to improve quality. There are also functional processes.

The output from a process is that which is transferred to somewhere or to someone – the *customer*. Clearly, to produce an output that meets the requirements of the customer, it is necessary to define, monitor and control the inputs to the process, which in turn may be supplied as output from an earlier process. At every supplier-customer interface then there resides a transformation process (Figure 1.4), and every single task throughout an organisation must be viewed as a process in this way.

Once we have established that our process is capable of meeting the requirements, we can address the next question, *'Do we continue to do the job correctly?'*, which brings a requirement to monitor the process and the controls on it. If we now re-examine the first question, 'Have we done the job correctly?', we can see that, if we have been able to answer the other two questions with a yes, we *must* have done the job correctly. Any other outcome would be illogical. By asking the questions in the right order, we have moved the need to ask the 'inspection' question and replaced a strategy of *detection* with one of *prevention*. This concentrates all the attention on the front end of any process – the inputs – and changes the emphasis to making sure the inputs are capable of meeting the requirements of the process. This is a managerial responsibility.

These ideas apply to every transformation process; they all must be subject to the same scrutiny of the methods, the people, skills, equipment and so on to make sure they are correct for the job. A person giving a lecture whose overhead projector equipment will not focus correctly, or whose teaching materials are not appropriate, will soon discover how difficult it is to provide a lecture that meets the requirements of the audience.

In every organisation there are some very large processes – groups of smaller processes called *key, critical or business processes*. These are activities the organisation must carry out especially well if its mission and objectives are to be achieved. The area

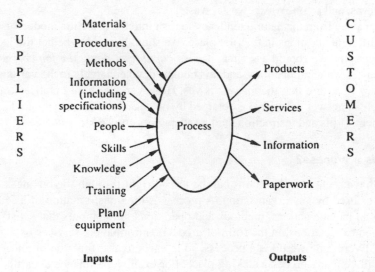

Figure 1.4   *A process*

will be dealt with in some detail in Chapter 13 on the implementation of TQM. It is crucial if the management of quality is to be integrated into the strategy for the organisation.

The *control* of quality clearly can only take place at the point of operation or production – where the letter is typed, the sales call made, the patient admitted, or the chemical manufactured. The act of *inspection is not quality control*. When the answer to 'Have we done the job correctly?' is given indirectly by answering the questions of capability and control, then we have *assured* quality, and the activity of checking becomes one of *quality assurance* – making sure that the product or service represents the output from an effective *system* to ensure capability and control. It is frequently found that organisational barriers between departmental empires encouraged the development of testing and checking of services or products in a vacuum, without interaction with other departments.

Quality control then is essentially the activities and techniques employed to achieve and maintain the quality of a product, process, or service. It includes a monitoring activity, but is also concerned with finding and eliminating causes of quality problems so that the requirements of the customer are continually met.

Quality assurance is broadly the prevention of quality problems through planned and systematic activities (including documentation). These will include the establishment of a good quality management system and the assessment of its adequacy, the audit of the operation of the system, and the review of the system itself.

1.4 Quality starts with 'marketing'

The authors have been asked on more than one occasion if TQM applies to marketing. The answer to the question is not remarkable – it starts there!

The marketing function of an organisation must take the lead in establishing the true requirements for the product or service. Having determined the need, marketing should define the market sector and demand, to determine such product or service features as grade, price, quality, timing, etc. For example, a major hotel chain thinking of opening a new hotel or refurbishing an old one will need to consider its location and accessibility before deciding whether it will be predominantly a budget, first-class, business or family hotel.

Marketing will also need to establish customer requirements by reviewing the market needs, particularly in terms of unclear or unstated expectations or preconceived ideas held by customers. Marketing is responsible for determining the key characteristics that determine the suitability of the product or service in the eyes of the customer. This may of course call for the use of market research techniques, data-gathering, and analysis of customer complaints. If possible, quasi-quantitative methods should be employed, giving proxy variables that can be used to grade the characteristics in importance, and decide in which areas superiority over competitors exists. It is often useful to compare these findings with internal perceptions of quality.

Excellent communication between customers and suppliers is the key to total quality; it will eradicate the 'demanding nuisance/idiot' view of customers, which pervades many organisations. Poor communications often occur in the supply chain between organisations, when neither party realises how poor they are. Feedback from both customers and suppliers needs to be improved where dissatisfied customers and suppliers do not communicate their problems. In such cases non-conformance of purchased products or services is often due to customers' inability to communicate their requirements clearly. If these ideas are also used within an organisation, then the internal supplier/customer interfaces will operate much more smoothly.

All the efforts devoted to finding the nature and timing of the demand will be pointless if marketing fails to communicate the requirements promptly, clearly, and accurately to the remainder of the organisation. The marketing function should be capable of supplying the company with a formal statement or outline of the requirements for each product or service. This constitutes a preliminary set of *specifications*, which can be used as the basis for service or product design. The information requirements include:

1   Characteristics of performance and reliability – these must make reference to the conditions of use and any environmental factors that may be important.
2   Aesthetic characteristics, such as style, colour, smell, taste, feel, etc.
3   Any obligatory regulations or standards governing the nature of the product or service.

Marketing must also establish systems for feedback of customer information and reaction, and these systems should be designed on a continuous monitoring basis. Any information pertinent to the product or service should be collected and collated, interpreted, analysed, and communicated, to improve the response to customer experience and expectations. These same principles must also be applied inside the organisation if continuous improvement at every transformation process interface is to be achieved. If one department of a company has problems recruiting the correct sort of staff, and personnel has not established mechanisms for gathering, analysing, and responding to information on new employees, then frustration and conflict will replace communication and co-operation.

One aspect of the analysis of market demand that extends back into the organisation is the review of market readiness of a new product or service. Items that require some attention include assessment of:

1   The suitability of the distribution and customer-service systems.
2   Training of personnel in the 'field'.
3   Availability of spare parts or support staff.
4   Evidence that the organisation is capable of meeting customer requirements.

All organisations receive a wide range of information from customers through invoices, payments, requests for information, letters of complaint, responses to advertisements and promotion, etc. An essential component of a system for the analysis of market demand is that this data is channelled quickly into the appropriate areas for action and, if necessary, response.

There are various techniques of market research, but they will not be described in detail in this book, for they are well documented elsewhere. Nevertheless it is worth listing some of the most common and useful general methods that should be considered for use, both externally and internally:

- Customer surveys.
- Quality panel or focus group techniques.
- In-depth interviews.
- Brainstorming and discussions.
- Role rehearsal and reversal.
- Interrogation of trade associations.

The number of methods and techniques for researching market demand is limited only by imagination and funds. The important point to stress is that the supplier, whether the internal individual or the external organisation, keeps very close to the customer. Market research, coupled with analysis of complaints data, is an essential part of finding out what the requirements are, and breaking out from the obsession with inward scrutiny that bedevils quality.

1.5 Quality in all functions

For an organisation to be truly effective, each part of it must work properly together. Each part, each activity, each person in the organisation affects and is in turn affected by others. Errors have a way of multiplying, and failure to meet the requirements in one part or area creates problems elsewhere, leading to yet more errors, yet more problems, and so on. The benefits of getting it right first time everywhere are enormous.

Everyone experiences – almost accepts – problems in working life. This causes people to spend a large part of their time on useless activities – correcting errors, looking for things, finding out why things are late, checking suspect information, rectifying and reworking, apologising to customers for mistakes, poor quality and lateness. The list is endless, and it is estimated that about one-third of our efforts are wasted in this way. In the service sector it can be much higher.

Quality, the way we have defined it as meeting the customer requirements, gives people in different functions of an organisation a common language for improvement. It enables all the people, with different abilities and priorities, to communicate readily with one another, in pursuit of a common goal. When business and industry were local, the craftsman could manage more or less on his own. Business is now so complex and employs so many different specialist skills that everyone has to rely on the activities of others in doing their jobs.

Some of the most exciting applications of TQM have materialised from departments that could see little relevance when first introduced to its concepts. Following training, many different departments of organisations can show the use of the techniques. Sales staff can monitor and increase successful sales calls, office staff have used TQM methods to prevent errors in word-processing and improve inputting to computers, customer-service people have monitored and reduced complaints, the distribution department has controlled lateness and disruption in deliveries.

It is worthy of mention that the first points of contact for some outside customers are the telephone operator, the security people at the gate, or the person in reception. Equally the paperwork and support services associated with the product, such as invoices and sales literature and their handlers, must match the needs of the customer. Clearly TQM cannot be restricted to the production or operational areas without losing great opportunities to gain maximum benefit.

Managements that rely heavily on exhortation of the workforce to 'do the right job right the first time', or 'accept that quality is your responsibility', will not only fail to achieve quality but will create division and conflict. These calls for improvement infer that faults are caused only by the workforce and that problems are departmental when, in fact, the opposite is true – most problems are inter-departmental. The commitment of all members of an organisation is a requirement of 'company-wide quality improvement'. Everyone must work together at every interface to achieve perfection. And that can only happen if the top management is really committed to quality improvement.

Chapter highlights

Quality and competitiveness

- The reputation enjoyed by an organisation is built by quality, reliability, delivery and price. Quality is the most important of these competitive weapons.
- Reputations for poor quality last for a long time, and good or bad reputations can become national. The management of quality can be learned and used to improve reputation.
- Quality is meeting the customer requirements, and this is not restricted to the functional characteristics of the product or service.
- Reliability is the ability of the product or service to continue to meet the customer requirements over time.
- Organisations 'delight' the customer by consistently meeting customer requirements, and then achieve a reputation of 'excellence'.

Understanding and building the quality chains

- Throughout all organisations there are a series of internal suppliers and customers. These form the so-called 'quality chains', the core of the company-wide quality improvement (CWQI).
- The internal customer/supplier relationships must be managed by interrogation, ie using a set of questions at every interface. Measurement of capability is vital.
- There are two distinct but interrelated aspects of quality, design and conformance to design. *Quality of design* is a measure of how well the product or service is designed to achieve the agreed requirements. *Quality of conformance to design* is the extent to which the product or service achieves the design. Organisations should assess how much time they spend doing the right things right.

Managing processes

- Asking the question 'Have we done the job correctly?' should be replaced by asking 'Are we capable of doing the job correctly?' and 'Do we continue to do the job correctly?'
- Asking the questions in the right order replaces a strategy of *detection* with one of *prevention*.

- Everything we do is a process, which is the transformation of a set of inputs into the desired outputs.
- In every organisation there are some key, critical or business processes that must be performed especially well if the mission and objectives are to be achieved.
- Inspection is not *quality control*. The latter is the employment of activities and techniques to achieve and maintain the quality of a product, process or service.
- *Quality assurance* is the prevention of quality problems through planned and systematic activities.

Quality starts with 'marketing'

- Marketing establishes the true requirements for the product or service. These must be communicated properly throughout the organisation in the form of specifications.

Quality in all functions

- All members of an organisation need to work together on 'company-wide quality improvement'. The co-operation of everyone at every interface is required to achieve perfection.

2
Commitment and leadership

2.1 The total quality management approach

'What is quality management?' Something that is best left to the experts is often the answer to this question. But this is avoiding the issue, because it allows executives and managers to opt out. Quality is too important to leave to the so called 'quality professionals'; it cannot be achieved on a company-wide basis if it is left to the experts. Equally dangerous, however, are the uninformed who try to follow their natural instincts because they 'know what quality is when they see it'. This type of intuitive approach will lead to serious attitude problems, which do no more than reflect the understanding and knowledge of quality that are present in an organisation.

The organisation which believes that the traditional quality control techniques, and the way they have always been used, will resolve their quality problems is wrong. Employing more inspectors, tightening up standards, developing correction, repair and rework teams do not promote quality. Traditionally, quality has been regarded as the responsibility of the QC department, and still it has not yet been recognised in some organisations that many quality problems originate in the service or administrative areas.

Total Quality Management is far more than shifting the responsibility of *detection* of problems from the customer to the producer. It requires a comprehensive approach that must first be recognised and then implemented if the rewards are to be realised. Today's business environment is such that managers must plan strategically to maintain a hold on market share, let alone increase it. We have known for years that consumers place a higher value on quality than on loyalty to home-based producers, and price is no longer the major determining factor in consumer choice. Price has been replaced by quality, and this is true in industrial, service, hospitality, and many other markets.

TQM is an approach to improving the competitiveness, effectiveness and flexibility of a whole organisation. It is essentially a way of planning, organising and understanding each activity, and depends on each individual at each level. For an organisation to be truly effective, each part of it must work properly together towards the same goals, recognising that each person and each activity affects and in turn is affected by others. TQM is also a way of ridding people's lives of wasted effort by bringing everyone into the processes of improvement, so that results are achieved in less time. The methods and techniques used in TQM can be applied throughout any organisation. They are equally useful in the manufacturing, public service, health care, education and hospitality industries. TQM needs to gain ground rapidly and become a way of life in many organisations.

The impact of TQM on an organisation is, firstly, to ensure that the management adopts a strategic overview of quality. The approach must focus on developing a *problem-prevention* mentality; but it is easy to underestimate the effort that is required to change attitudes and approaches. Many people will need to undergo a complete change of 'mindset' to unscramble their intuition, which rushes into the detection/inspection mode to solve quality problems – 'We have a quality problem, we had better check every letter – take two samples out of each sack – check every widget twice', etc.

The correct mindset may be achieved by looking at the sort of barriers that exist in key areas. Staff will need to be trained and shown how to reallocate their time and energy to studying their processes in teams, searching for causes of problems, and correcting the causes, not the symptoms, once and for all. This will require of management a positive, thrusting initiative to promote the right-first-time approach to work situations. Through *quality improvement teams*, which will need to be set up, these actions will reduce the inspection-rejection syndrome in due course. If things are done correctly first time round, the usual problems that create the need for inspection for failure will disappear.

The managements of many firms may think that their scale of operation is not sufficiently large, that their resources are too slim, or that the need for action is not important enough to justify implementing TQM. Before arriving at such a conclusion, however, they should examine the existing quality performance by asking the following questions:

1   Is any attempt made to assess the costs arising from errors, defects, waste, customer complaints, lost sales, etc? If so, are these costs minimal or insignificant?
2   Is the standard of quality management adequate and are attempts being made to ensure that quality is given proper consideration at the design stage?
3   Are the organisation's quality systems – documentation, procedures, operations etc – in good order?
4   Have personnel been trained in how to prevent errors and quality problems? Do they anticipate and correct potential causes of problems, or do they find and reject?
5   Do job instructions contain the necessary quality elements, are they kept up-to-date, and are employers doing their work in accordance with them?
6   What is being done to motivate and train employees to do work right first time?
7   How many errors and defects and how much wastage occurred last year? Is this more or less than the previous year?

If satisfactory answers can be given to most of these questions, an organisation can be reassured that it is already well on the way to using adequate quality procedures and management. Even so, it may find that the introduction of TQM causes it to reappraise quality activities throughout. If answers to the above questions indicate problem areas, it will be beneficial to review the top management's attitude to quality. Time and money spent on quality-related activities are *not* limitations of profitability; they make significant contributions towards greater efficiency and enhanced profits.

2.2 Commitment and policy

To be successful in promoting business efficiency and effectiveness, TQM must be truly organisation-wide, and it must start at the top with the Chief Executive or equivalent. The most senior directors and management must all demonstrate that they are serious about quality. The middle management have a particularly important role to play, since they must not only grasp the principles of TQM, they must go on to explain them to the people for whom they are responsible, and ensure that their own commitment is communicated. Only then will TQM spread effectively throughout the organisation. This level of management must also ensure that the efforts and achievements of their subordinates obtain the recognition, attention and reward that they deserve.

The Chief Executive of an organisation must accept the responsibility for and commitment to a quality policy in which he/she must really believe. This commitment is part of a broad approach extending well beyond the accepted formalities of the quality assurance function. It creates responsibilities for a chain of quality interactions between the marketing, design, production/operations, purchasing, distribution and service functions. Within each and every department of the organisation at all levels, starting at the top, basic changes of attitude will be required to operate TQM. If the owners or directors of the organisation do not recognise and accept their responsibilities for the initiation and operation of TQM, then these changes will not happen. Controls, systems and techniques are very important in TQM, but they are not the primary requirement. It is more an attitude of mind, based on pride in the job and teamwork, and it requires from the management total commitment, which must then be extended to all employees at all levels and in all departments.

Senior management commitment must be obsessional, not lip service. It is possible to detect real commitment; it shows on the shop floor, in the offices, in the hospital ward – at the point of operation. Going into organisations sporting poster-campaigning for quality instead of belief, one is quickly able to detect the falseness. The people are told not to worry if quality problems arise, 'just do the best you can', 'the customer will never notice'. The opposite is an organisation where total quality means something, can be seen, heard, felt. Things happen at this operating interface as a result of *real* commitment. Material problems are corrected with suppliers, equipment difficulties are put right by improved maintenance programmes or replacement, people are trained, change takes place, partnerships are built, continuous improvement is achieved.

The quality policy

A sound quality policy, together with the organisation and facilities to put it into effect, is a fundamental requirement, if a company is to begin to implement TQM. Every organisation should develop and state its policy on quality, together with arrangements for its implementation. The contents of the policy should be made known to all employees. The preparation and implementation of a properly thought out quality policy, together with continuous monitoring, make for smoother production or service operation, minimise errors and reduce waste.

Management must be dedicated to the regular improvement of quality, not simply a one-step improvement to an acceptable plateau. These ideas must be set out in a *quality policy* that requires top management to:

1 Establish an 'organisation' for quality.
2 Identify the customer's needs and perception of needs.
3 Assess the ability of the organisation to meet these needs economically.
4 Ensure that bought-in materials and services reliably meet the required standards of performance and efficiency.
5 Concentrate on the prevention rather than detection philosophy.
6 Educate and train for quality improvement.
7 Review the quality management systems to maintain progress.

The quality policy must be publicised and understood at all levels of the organisation.

An example of a good company quality policy is given below:

- Quality improvement is primarily the responsibility of management.
- In order to involve everyone in the organisation in quality improvement, management will enable all employees to participate in the preparation, implementation and evaluation of improvement activities.
- Quality improvement will be tackled and followed up in a systematic and planned manner. This applies to every part of our organisation.
- Quality improvement will be a continuous process.
- The organisation will concentrate on its customers and suppliers, both external and internal.
- The performance of our competitors will be shown to all relevant units.
- Important suppliers will be closely involved in our quality policy. This relates to both external and internal suppliers of goods, resources, and services.
- Widespread attention will be given to education and training activities, which will be assessed with regard to their contribution to the quality policy.
- Publicity will be given to the quality policy in every part of the organisation so that everyone may understand it. All available methods and media will be used for its internal and external promotion and communication.
- Reporting on the progress of the implementation of the policy will be a permanent agenda item in management meetings.

The quality policy must be the concern of all employees, and the principles and objectives communicated as widely as possible. Practical assistance and training should be given, where necessary, to ensure the relevant knowledge and experience are acquired for successful implementation of the policy.

2.3 Creating or changing the culture

The culture within an organisation is formed by a number of components:
1 Behaviours based on people interactions.
2 Norms resulting from working groups.
3 Dominant values adopted by the organisation.
4 Rules of the game for getting on.
5 The climate.

Culture in any 'business' may be defined then as the beliefs that pervade the organisation about how business should be conducted, and how employees should behave and should be treated. Any organisation needs a vision framework that includes its *guiding philosophy, core values and beliefs* and a *purpose*. These should be combined into a *mission*, which provides a vivid description of what things will be like when it has been achieved.

The *guiding philosophy* drives the organisation and is shaped by the leaders through their thoughts and actions. It should reflect the vision of an organisation rather than the vision of a single leader, and should evolve with time, although organisations must hold on to the *core* elements.

The *core values and beliefs* represent the organisation's basic principles about what is important in business, its conduct, its social responsibility and its response to changes in the environment. They should act as a guiding force, with clear and authentic values, which are focused on employees, suppliers, customers, society at large, safety, shareholders, and generally stakeholders.

The *purpose* of the organisation should be a development from the core values and beliefs and should quickly and clearly convey how the organisation is to fulfil its role.

The *mission* will translate the abstractness of philosophy into tangible goals that will move the organisation forward and make it perform to its optimum. It should not be limited by the constraints of strategic analysis, and should be proactive not reactive. Strategy is subservient to mission, the strategic analysis being done after, not during, the mission setting process.

Control

The effectiveness of an organisation and its people depends on the extent to which each person and department perform their role and move towards the common goals and objectives. Control is the process by which information or feedback is provided so as to keep all functions on track. It is the sum total of the activities that increase the probability of the planned results being achieved. Control mechanisms fall into three categories, depending upon their position in the managerial process:

Before the fact	*Operational*	*After the fact*
Strategic plan	Observation	Annual reports
Action plans	Inspection and correction	Variance reports
Budgets	Progress review	Audits
Job descriptions	Staff meetings	Surveys
Individual performance-objectives	Internal Information and data systems	Performance Review
Training and development plans	Training programmes	Evaluation of training

Many organisations use after-the-fact controls, causing managers to take a reactive rather than a proactive position. Such 'crisis-orientation' needs to be replaced by a more anticipative one in which the focus is on preventive or before-the-fact controls.

Attempting to control performance through systems, procedures, or techniques *external* to the individual is not an effective approach, since it relies on 'controlling' others;

individuals should be responsible for their own actions. An externally based control system can result in a high degree of concentrated effort in a specific area if the system is overly structured, but it can also cause negative consequences to surface:

1   Since all rewards are based on external measures, which are imposed, the 'team members' often focus all their efforts on the measure itself, eg to have it set lower (or higher) than possible, to manipulate the information which serves to monitor it, or to dismiss it as someone else's goal not theirs. In the budgeting process, for example, distorted figures are often submitted by those who have learned that their 'honest projections' will be automatically altered anyway.
2   When the rewards are dependent on only one or two limited targets, all efforts are directed at those, even at the expense of others. If short-term profitability is the sole criterion for bonus distribution or promotion, it is likely that investment for longer-term growth areas will be substantially reduced. Similarly, strong emphasis and reward for output or production may result in lowered quality.
3   The fear of not being rewarded, or even being criticised, for performance that is less than desirable may cause some to withhold information that is unfavourable but nevertheless should be flowing into the system.
4   When reward and punishment are used to motivate performance, the degree of risk-taking may lessen and be replaced by a more cautious and conservative approach. In essence, the fear of failure replaces the desire to achieve.

The following problem situations have been observed by the authors and their colleagues within companies that have taken part in research and consultancy on quality management:

•   The goals imposed are seen or known to be unrealistic. If the goals perceived by the subordinate are in fact accomplished, then the subordinate has proved himself wrong. This clearly has a negative effect on the effort expended, since few people are motivated to prove themselves wrong!
•   Where individuals are stimulated to commit themselves to a goal, and where their personal pride and self-esteem are at stake, then the level of motivation is at a peak. For most people the toughest critic and the hardest taskmaster they confront is not their immediate boss but themselves.
•   Directors and managers are often afraid of allowing subordinates to set the goals for fear of them being set too low, or loss of control over subordinate behaviour. It is also true that many do not wish to set their own targets, but prefer to be told what is to be accomplished.

TQM is concerned with moving the focus of control from outside the individual to within, the objective being to make everyone accountable for their own performance, and to get them committed to attaining quality in a highly motivated fashion. The assumptions a director or manager must make in order to move in this direction are simply that people do not need to be coerced to perform well, and that people want to achieve, accomplish, influence activity, and challenge their abilities. If there is belief in this, then only the techniques remain to be discussed.

Total Quality Management is user-driven – it cannot be imposed from outside the organisation, as perhaps can a quality standard or statistical process control. This means that the ideas for improvement must come from those with knowledge and experience of the processes, activities and tasks; this has massive implications for training and follow-up. TQM is not a cost-cutting or productivity improvement device in the traditional sense, and it must not be used as such. Although the effects of a successful programme will certainly reduce costs and improve productivity, TQM is concerned chiefly with changing attitudes and skills so that the culture of the organisation becomes one of preventing failure – doing the right things, right first time, every time.

2.4 Effective leadership

Some management teams have broken away from the traditional style of management; they have made a 'managerial breakthrough'. Their new approach puts their organisations head and shoulders above competitors in the fight for sales, profits, resources, funding and jobs. Many service organisations are beginning to move in the same way, and the successful quality-based strategy they are adopting depends very much on effective leadership.

Effective leadership starts with the Chief Executive's vision, capitalising on market or service opportunities, continues through a strategy that will give the organisation competitive advantage, and leads to business or service success. It goes on to embrace all the beliefs and values held, the decisions taken and the plans made by anyone anywhere in the organisation, and the focusing of them into effective, value-adding action.

Together, effective leadership and total quality management result in the company or organisation doing the right things, right first time.

The five requirements for effective leadership are the following.

1 Developing and publishing clear documented corporate beliefs and objectives – a mission statement

Executives must express values and beliefs through a clear vision of what they want their company or organisation to be, and through objectives – what they specifically want to achieve in line with the basic beliefs. Together, they define what the company or organisation is all about. The senior management team will need to spend some time away from the 'coal face' to do this and develop their programme for implementation.

Clearly defined and properly communicated beliefs and objectives, which can be summarised in the form of a mission statement, are essential if the directors, managers and other employees are to work together as a winning team. The beliefs and objectives should address:

- The definition of the business, eg the needs that are satisfied or the benefits provided.
- A commitment to effective leadership and quality.
- Target sectors and relationships with customers, and market or service position.
- The role or contribution of the company, organisation, or unit, eg profit-generator, service department, opportunity-seeker.

- The distinctive competence – a brief statement which applies only to that organisation, company or unit.
- Indications for future direction – a brief statement of the principal plans which would be considered.
- Commitment to monitoring performance against customers' needs and expectations, and continuous improvement.

The mission statement and the broad beliefs and objectives may then be used to communicate an inspiring vision of the organisation's future. The top management must then show *TOTAL COMMITMENT* to it.

2 *Developing clear and effective strategies and supporting plans for achieving the mission and objectives*

The achievement of the company or service objectives requires the development of business or service strategies, including the strategic positioning in the 'market place'. Plans for implementing the strategies can then be developed. Strategies and plans can be developed by senior managers alone, but there is likely to be more commitment to them if employee participation in their development and implementation is encouraged.

3 *Identifying the critical success factors and critical processes*

The next step is the identification of the *critical success factors* (CSFs), a term used to mean the most important subgoals of a business or organisation. CSFs are what must be accomplished for the mission to be achieved. The CSFs are followed by the key, critical or business processors for the organisation – the activities that must be done particularly well for the CSFs to be achieved. This process is described in some detail in Chapter 13 on implementation.

4 *Reviewing the management structure*

Defining the corporate objectives and strategies, CSFs and critical processes might make it necessary to review the organisational structure. Directors, managers and other employees can be fully effective only if an effective structure based on process management exists. This includes both the definition of responsibilities for the organisation's management and the operational procedures they will use. These must be the agreed best ways of carrying out the critical processes.

The review of the management structure should include the establishment of a process quality improvement team structure throughout the organisation.

5 *Empowerment – encouraging effective employee participation*

For effective leadership it is necessary for management to get very close to the employees. They must develop effective communications – up, down and across the organisation – and take action on what is communicated; and they must encourage good communications between all suppliers and customers.

Particular attention must be paid to the following.

Attitudes

The key attitude for managing any winning company or organisation may be expressed as follows: 'I will personally understand who my customers are and what are their needs and expectations of me; I will measure how well I am satisfying their needs and expectations and I will take whatever action is necessary to satisfy them fully. I will also understand and communicate my requirements to my suppliers, inform them of changes and provide feedback on their performance'. This attitude must start at the top – with the Chairman or Chief Executive. It must then percolate down, to be adopted by each and every employee. That will happen only if managers lead by example. Words are cheap and will be meaningless if employees see from managers' actions that they do not actually believe or intend what they say.

Abilities

Every employee must be able to do what is needed and expected of him or her, but it is first necessary to decide what is really needed and expected. If it is not clear what the employees are required to do and what standards of performance are expected, how can managers expect them to do it?

Train, train, train and train again. Training is very important, but it can be expensive if the money is not spent wisely. The training must be related to needs, expectations, and process improvement. It must be planned and *always* its effectiveness must be reviewed.

Participation

If all employees are to participate in making the company or organisation successful (directors and managers included), then they must also be trained in the basics of disciplined management.

They must be trained to:

E Evaluate – the situation and define their objectives.
P Plan – to achieve those objectives fully.
D Do, ie implement the plans.
C Check – that the objectives are being achieved.
A Amend, ie take corrective action if they are not.

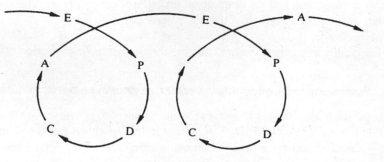

Figure 2.1   *The helix of never-ending improvement*

The word 'disciplined' applied to people at all levels means that they will do what they say they will do. It also means that in whatever they do they will go through the full process of Evaluate, Plan, Do, Check and Amend, rather than the more traditional and easier option of starting by doing rather than evaluating. This will lead to a never-ending improvement helix (Figure 2.1).

This basic approach needs to be backed up with good project management, planning techniques and problem-solving methods, which can be taught to anyone in a relatively short period of time. The project management enables changes to be made successfully and the problem-solving helps people to remove the obstacles in their way. Directors and managers need this training as much as other employees.

2.5 Ten points for senior management – the foundations of the TQM model

The vehicle for achieving effective leadership is Total Quality Management. We have seen that it covers the entire organisation, all the people and all the functions, including external organisations and suppliers. In the first two chapters, several facets of TQM have been reviewed, including:

- Recognising customers and discovering their needs.
- Setting standards that are consistent with customer requirements.
- Controlling processes, including systems, and improving their capability.
- Management's responsibility for setting the guiding philosophy, quality policy, etc, and providing motivation through leadership and equipping people to achieve quality.
- Empowerment of people at all levels in the organisation to act for quality improvement.

The task of implementing TQM can be daunting, and the Chief Executive and directors faced with it may become confused and irritated by the proliferation of theories and packages. A simplification is required. The *core* of TQM must be the customer-supplier interfaces, both internally and externally, and the fact that at each interface there are processes to convert inputs to outputs. Clearly, there must be commitment to building-in quality through management of the inputs and processes.

How can senior managers and directors be helped in their understanding of what needs to be done to become committed to quality and implement the vision? Some American and Japanese quality 'gurus' have each set down a number of points or absolutes – words of wisdom in management and leadership – and many organisations are using these to establish a policy based on quality. These have been distilled down and modified here to ten points for senior management to adopt.

1   The organisation needs long term COMMITMENT to constant improvement

There must be a constancy of purpose, and commitment to it must start from the top. The quality improvement process must be planned on a truly organisation-wide basis, ie it must embrace all locations and departments and must include customers, suppliers, and subcontractors. It cannot start in 'one department' in the hope that the programme will spread from there.

The place to start the quality process is in the boardroom – leadership must be by example. Then the process must *progressively* expand to embrace all parts of the organisation. It is wise to avoid the 'blitz' approach to TQM implementation, for it can lead to a lot of hype but no real changes in behaviour.

2   Adopt the philosophy of zero errors/defects to change the CULTURE to right first time

This must be based on a thorough understanding of the customer's needs and expectations, and on teamwork, developed through employee participation and rigorous application of the EPDCA helix.

3   Train the people to understand the CUSTOMER–SUPPLIER relationships

Again the commitment to customer needs must start from the top, from the Chairman or Chief Executive. Without that, time and effort will be wasted. Customer orientation must then be achieved for each and every employee, directors and managers. The concept of internal customers and suppliers must be thoroughly understood and used.

4   Do not buy products or services on price alone – look at the TOTAL COST

Demand continuous improvement in everything, including suppliers. This will bring about improvements in product, service and failure rates. Continually improve the product or the service provided externally, so that the total costs of doing business are reduced.

5   Recognise that improvement of the SYSTEMS needs to be managed

Defining the performance standards expected and the systems to achieve them is a managerial responsibility. The rule has to be that the systems will be in line with the shared needs and expectations and will be part of the continuous improvement process.

6   Adopt modern methods of SUPERVISION and TRAINING – eliminate fear

It is all too easy to criticise mistakes, but it often seems difficult to praise efforts and achievements. Recognise and publicise efforts and achievements and provide the right sort of training, facilitation and supervision.

7   Eliminate barriers between departments by managing the PROCESS – improve COMMUNICATIONS and TEAMWORK

Barriers are often created by 'silo management', in which departments are treated like containers that are separate from one another. The customers are not interested in departments; they stand outside the organisation and see slices through it – the *processes*. It is necessary to build teams and improve communications around the processes.

8   Eliminate the following:

- Arbitrary goals without methods.
- All standards based only on numbers.
- Barriers to pride of workmanship.
- Fiction. Get *FACTS* by using the correct *TOOLS*.

At all times it is essential to know how well you are doing in terms of satisfying the customers' needs and expectations. Help all employees to know *how* they will achieve their goals and how well they are doing.

Traditional piecework will not survive in a TQM environment, or *vice-versa*, because it creates barriers and conflict. People should be proud of what they do and not be encouraged to behave like monkeys being thrown peanuts.

Train people to measure and report performance in language that the people doing the job can understand. Encourage each employee to measure his/her own performance. Do not stop with measuring performance in the organisation – find out how well other organisations (competitive or otherwise) are performing against similar needs and expectations (*benchmark* against best practice).

The costs of quality mismanagement and the level of firefighting are excellent factual indicators of the internal health of an organisation. They are relatively easily measured and simple for most people to understand.

9   Constantly educate and retrain – develop the 'EXPERTS' in the business

The experts in any business are the people who do the job every day of their lives. The 'energy' that lies within them can be released into the organisation through education, training, encouragement and the chance to participate.

10   Develop a SYSTEMATIC approach to manage the implementation of TQM

TQM should not be regarded as a woolly-minded approach to running an organisation. It requires a carefully planned and fully integrated strategy, derived from the mission. That way it will help any organisation to realise its vision.

Summary

- Identify *customer–supplier* relationships.
- Manage *processes*.
- Change the *culture*.
- Improve *communication*.
- Show *commitment*.

The right culture, communication, and commitment form the basis of the first part of a model for TQM – the 'soft' outcomes of TQM (Figure 2.2). The process core must be surrounded, however, by some 'hard' management necessities:

1   Systems (based on a good international standard, see Part 2 of this book).
2   Tools (for analysis, correlations, and predictions for action for continuous improvement to be taken, see Part 3 of this book).
3   Teams (the councils, quality improvement teams, quality circles, corrective action teams, etc, see Part 4 of this book).

The model now provides a multi-dimensional TQM 'vision' against which a particular organisation's status can be examined, or against which a particular approach to TQM implementation may be compared and weaknesses highlighted. It is difficult to draw in only two dimensions, but Figure 2.3 is an attempt to represent the major features of the model, the implementation of which is dealt with in Part 5.

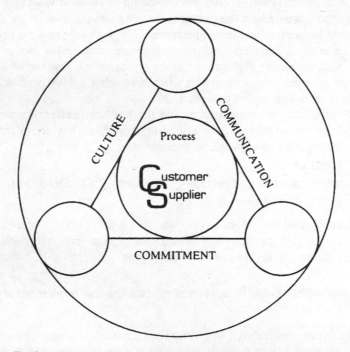

Figure 2.2   *Total quality management model – the 'soft' outcomes*

One of the greatest tangible benefits of improved quality is the increased market share that results, rather than just the reduction in quality costs. The evidence for this can be seen already in some of the major consumer and industrial markets of the world. Superior quality can also be converted into premium prices. Quality clearly correlates with profit. The less tangible benefit of greater employee participation in quality is equally, if not more, important in the longer term. The pursuit of continual improvement must become a way of life for everyone in an organisation if it is to succeed in today's competitive environment.

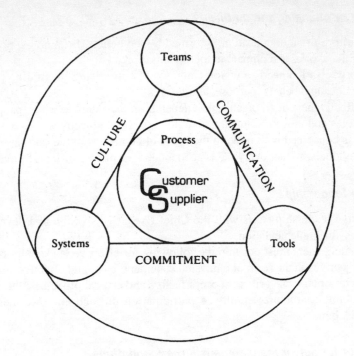

Figure 2.3   *Total quality management model – major features*

Chapter highlights

The Total Quality Management approach

- TQM is a comprehensive approach to improving competitiveness, effectiveness and flexibility through planning, organising and understanding each activity, and involving each individual at each level. It is useful in all types of organisation.
- TQM ensures that management adopts a strategic overview of quality and focuses on prevention, not detection, of problems.
- It often requires a mindset change to break down existing barriers. Managements that doubt the applicability of TQM should ask questions about the operation's costs, errors, wastes, standards, systems, training and job instructions.

Commitment and policy

- TQM starts at the top, where serious obsessional commitment to quality must be demonstrated. Middle management also has a key role to play in communicating the message.
- Every Chief Executive must accept the responsibility for commitment to a quality policy that deals with the organisation for quality, the customer needs, the ability of the organisation, supplied materials and services, education and training, and review of the management systems for never-ending improvement.

Creating or changing the culture

*   The culture of an organisation is formed by the beliefs, behaviours, norms, dominant values, rules and climate in the organisation.
*   Any organisation needs a vision framework, comprising its guiding philosophy, core values and beliefs, purpose, and mission.
*   The effectiveness of an organisation depends on the extent to which people perform their roles and move towards the common goals and objectives.
*   TQM is concerned with moving the focus of control from the outside to the inside of individuals, so that everyone is accountable for his/her own performance.

Effective leadership

*   Effective leadership starts with the Chief Executive's vision and develops into a strategy for implementation.
*   Top management must develop the following for effective leadership: clear beliefs and objectives in the form of a mission statement; clear and effective strategies and supporting plans; the critical success factors and critical processes; the appropriate management structure; employee participation through empowerment, and the EPDCA helix.

Ten points for senior management – the foundations

*   Total quality is the key to effective leadership through commitment to constant improvement, a right first time philosophy, training people to understand customer-supplier relationships, not buying on price alone, managing systems improvement, modern supervision and training, managing processes through teamwork and improved communications, elimination of barriers and fear, constant education and 'expert' development, a systematic approach to TQM implementation.
*   The core of TQM is the customer-supplier relationship, where the processes must be managed. The 'soft' outcomes of TQM – the culture, communications, and commitment provide the foundation for the TQM model.
*   The process core must be surrounded by the 'hard' management necessities of systems, tools and teams. The model provides a framework against which an organisation's progress towards TQM can be examined.

3

Design for quality

3.1 Innovation, design and improvement

All businesses competing on the basis of quality need to update their products, processes and services periodically. In markets such as electronics, audio and visual goods, and office automation, new variants of products are offered frequently – almost like fashion goods. While in other markets the pace of innovation may not be as fast and furious, there is no doubt that the rate of change for product, service and process design has accelerated on a broad front.

Innovation entails both the invention and design of radically new products and services, embodying novel ideas, discoveries and advanced technologies, *and* the continuous development and improvement of existing products, services, and processes to enhance their performance and quality. It may also be directed at reducing costs of production or operations throughout the life cycle of the product or service system.

In many organisations innovation is predominantly either technology-led, eg in some chemical and engineering industries, or marketing-led, eg in some food companies. What is always striking about leading product or service innovators is that their developments are market-led, which is different from marketing-led. The latter means that the marketing function takes the lead in product and service developments. But most leading innovators identify and set out to meet the existing and potential demands profitably, and therefore are market-led, constantly striving to meet the requirements even more effectively through appropriate experimentation.

Commitment to quality in the most senior management helps to build quality throughout the design process and to ensure good relationships and communication between various groups and functional areas. Designing customer satisfaction into products and services contributes greatly to competitive success. Clearly, it does not guarantee it, because the conformance aspect of quality must be present and the operational processes must be capable of producing to the design. As in the marketing/operations interfaces, it is never acceptable to design a product, service, system or process that the customer wants but the organisation is incapable of achieving.

The design process, then, often concerns technological innovation in response to, or in anticipation of, changing market requirements and trends in technology. Those companies with impressive records of product- or service-led growth have demonstrated a state-of-the-art approach to innovation based on three principles:

- *Strategic balance* to ensure that both old and new product service developments are important. Updating old products, services and processes, ensures continuing cash generation from which completely new products may be funded.
- *Top management approach* to design to set the tone and ensure that commitment is the common objective by visibly supporting the design effort. Direct control should be concentrated on critical decision points, since over-meddling by very senior people in day-to-day project management can delay and demotivate staff.
- *Teamwork*, to ensure that once projects are under way, specialist inputs, eg from marketing and technical experts, are fused and problems are tackled simultaneously. The teamwork should be urgent yet informal, for too much formality will stifle initiative, flair and fun of design.

The extent of the design activity should not be underestimated, but it often is. Many people associate design with *styling* of products, and this is certainly an important aspect. But for certain products and many service operations the *secondary design* considerations are vital. Anyone who has bought an 'assemble-it-yourself' kitchen unit will know the importance of the design of the assembly instructions, for example. Aspects of design that affect quality in this way are packaging, customer-service arrangements, maintenance routines, warranty details and their fulfilment, spare-part availability, etc.

An industry that has learned much about the secondary design features of its products is personal computers. Many of the problems of customer dissatisfaction experienced in this market have not been product design features but problems with user manuals, availability and loading of software, and applications. For technically complex products or service systems, the design and marketing of after-sales arrangements are an essential component of the design activity. The design of production equipment and its layout to allow ease of access for repair and essential maintenance, or simple use as intended, widens the management of design quality into suppliers and contractors and requires their total commitment.

Proper design of plant and equipment plays a major role in the elimination of errors, defectives, and waste. Correct initial design also obviates the need for costly and wasteful modifications to be carried out after the plant or equipment has been constructed. It is at the plant design stage that such important matters as variability, reproducibility, ease or use in operation, maintainability, etc should receive detailed consideration.

Designing

If quality design is taking care of all aspects of the customer's requirements, including cost, production, safe and easy use, and maintainability of products and services, then *designing* must take place in all aspects of:

- Identifying the need (including need for change).
- Developing that which satisfies the need.
- Checking the conformance to the need.
- Ensuring that the need is satisfied.

Designing covers every aspect, from the identification of a problem to be solved, usually a market need, through the development of design concepts and prototypes to the generation of detailed specifications or instructions required to produce the artefact or provide the service. It is the process of presenting needs in some physical form, initially as a solution, and then as a specific configuration or arrangement of materials, resources, equipment, and people.

3.2 Quality function deployment (QFD) – the house of quality

The 'house of quality' is the framework of the approach to design management known as quality function deployment (QFD). It originated in Japan in 1972 at Mitsubishi's Kobe shipyard, but it has been developed in numerous ways by Toyota and its suppliers, and many other organisations. The house of quality (HOQ) concept, initially referred to as quality tables, has been used successfully by manufacturers of integrated circuits, synthetic rubber, construction equipment, engines, home appliances, clothing, and electronics, mostly Japanese. Ford and General Motors use it, and other organisations, including AT&T, Bell Laboratories, Digital Equipment, Hewlett-Packard, Procter & Gamble, ITT, Rank Xerox, Jaguar, and Mercury have applications. In Japan its design applications include public services, retail outlets, and apartment layout.

Quality function deployment (QFD) is a 'system' for designing a product or service, based on customer demands, with the participation of members of all functions of the supplier organisation. It translates the customer's requirements into the appropriate technical requirements for each stage. The activities included in QFD are:

1   Market research.
2   Basic research.
3   Invention.
4   Concept design.
5   Prototype testing.
6   Final-product or service testing.
7   After-sales service and trouble-shooting.

These are performed by people with different skills in a team whose composition depends on many factors, including the products or services being developed and the size of the operation. In many customer industries, such as cars, video equipment, electronics, and computers, 'engineering' designers are seen to be heavily into designing. But in other industries and service operations designing is carried out by people who do not carry the word 'designer' in their job title. The failure to recognise the design inputs they make, and to provide appropriate training and support, will limit the success of the design activities and result in some offering that does not satisfy the customer. This is particularly true of internal customers.

The QFD team in operation

The first step of a QFD exercise is to form a cross-functional QFD team. Its purpose is to take the needs of the market and translate them into such a form that they can be satisfied within the operating unit and delivered to the customers.

As with all organisational problems, the structure of the QFD Team must be decided on the basis of the detailed requirements of each organisation. One thing, however, is clear – close liaison must be maintained at all times between the design, marketing and operational functions represented in the team.

The QFD team must answer three questions – WHO, WHAT and HOW, ie

WHO are the customers?
WHAT does the customer need?
HOW will the needs be satisfied?

WHO may be decided by asking 'Who will benefit from the successful introduction of this product, service, or process?' Once the customers have been identified, WHAT can be ascertained through an interview/questionnaire process, or from the knowledge and judgement of the QFD team members. HOW is more difficult to determine, and will consist of the attributes of the product, service, or process under development. This will constitute many of the action steps in a 'QFD strategic plan'.

WHO, WHAT, and HOW are entered into the QFD matrix or grid of 'house of quality', which is a simple 'quality table'. The *WHAT*s are recorded in rows and the *HOW*s are placed in the columns.

The house of quality provides structure to the design and development cycle, often likened to the construction of a house, because of the shape of matrices when they are fitted together. The key to building the house is the focus on the customer requirements, so that the design and development processes are driven more by what the customer needs than by innovations in technology. This ensures that more effort is used to obtain the vital customer information. It may increase the initial planning time in a particular development project, but the time, including design and redesign, taken to bringing a product or service to the market will be reduced.

This requires that marketing people, design staff (including engineers), and production/operations personnel work closely together from the time the new service, process, or product is conceived. It will need to replace in many organisations the 'throwing it over the wall' approach, where a solid wall exists between each pair of functions (Figure 3.1).

The HOQ provides an organisation with the means for inter-departmental or inter-functional planning and communications, starting with the so-called customer attributes (CAs). These are phrases customers use to describe product, process, and service characteristics.

A complete QFD project will lead to the construction of a sequence of house of quality diagrams, which translate the customer requirements into specific operational process steps. For example, the 'feel' that customers like on the steering wheel of a motor car may translate into a specification for 45 standard degrees of synthetic polymer hardness, which in turn translates into specific manufacturing process steps, including the use of certain catalysts, temperatures, processes, and additives.

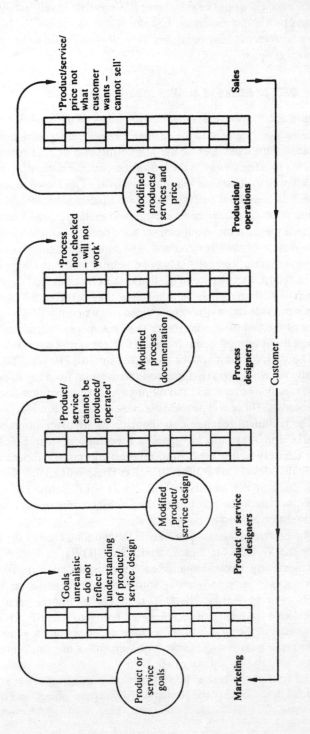

Figure 3.1 *'Throw it over the wall'. The design and development process is sequential and walled into separate functions*

The first steps in QFD lead to a consideration of the product as a whole, and subsequent steps to consideration of the individual components. For example, a complete hotel service would be considered at the first level, but subsequent QFD exercises would tackle the restaurant, bedrooms and reception. Each of the sub-services would have customer requirements, but they all would need to be compatible with the general service concept.

The QFD or house of quality tables

Figure 3.2 shows the essential components of the quality table or HOQ diagram. The construction begins with the *customer requirements*, which are determined through the 'voice of the customer' – the marketing and market research activities. These are entered into the blocks to the left of the central relationship matrix. Understanding and prioritising the customer requirements by the QFD team may require the use of competitive and compliant analysis, focus groups, and the analysis of market potential. The prime or broad requirements should lead to the detailed WHATs.

Once the customer requirements have been determined and entered into the table, the *importance* of each is rated and rankings are added. The use of the 'emphasis technique' or paired comparison may be helpful here (see Chapter 8).

Each customer requirement should then be examined in terms of customer rating; a group of customers may be asked how they perceive the performance of the organisation's product or service versus those of competitors'. These results are placed to the right of the central matrix. Hence the customer requirements' importance rankings and competition ratings appear from left to right across the house.

The WHATs must now be converted into the HOWs. These are called the *technical design requirements* and appear on the diagram from top to bottom in terms of requirements, rankings (or costs) and ratings against competition (technical benchmarking, see Chapter 7). These will provide the 'voice of the process'.

The technical requirements themselves are placed immediately above the central matrix and may also be given a hierarchy of prime and detailed requirements. Immediately below the central relationship matrix appear the rankings of technical difficulty, development time, or costs. These will enable the QFD team to discuss the efficiency of the various technical solutions. Below the technical rankings on the diagram comes the benchmark data, which compares the technical processes of the organisation against its competitors'.

The *central relationship matrix* is the working core of the house of quality diagram. Here the WHATs are matched with the HOWs, and each customer requirement is systematically assessed against each technical design requirement. The nature of any relationship – strong positive, positive, neutral, negative, strong negative – is shown by symbols in the matrix. The QFD team carries out the relationship estimation, using experience and judgement, the aim being to identify HOW the WHATs may be achieved. All the HOWs listed must be necessary and together sufficient to achieve the WHATs. Blank rows (customer requirement not met) and columns (redundant technical characteristics) should not exist.

The roof of the house shows the interactions between the technical design requirements. Each characteristic is matched against the others, and the diagonal format allows

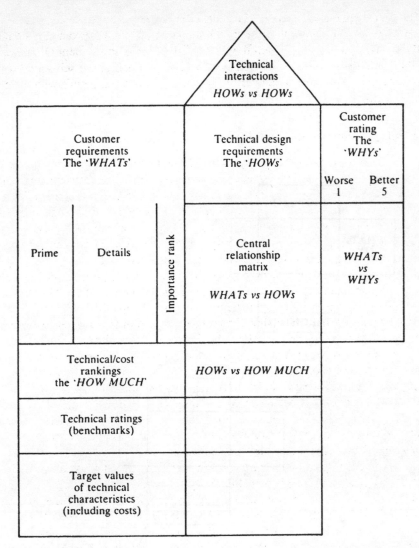

Figure 3.2 *The house of quality*

the nature of relationships to be displayed. The symbols used are the same as those in the central matrix.

The complete QFD process is time-consuming, because each cell in the central and roof matrices must be examined by the whole team. The team must examine the matrix to determine which technical requirement will need design attention, and the costs of that attention will be given in the bottom row. If certain technical costs become a major issue, the priorities may then be changed. It will be clear from the central matrix if there is more than one way to achieve a particular customer requirement, and the roof matrix will show if the technical requirements to achieve one customer requirement will have a negative effect on another technical issue.

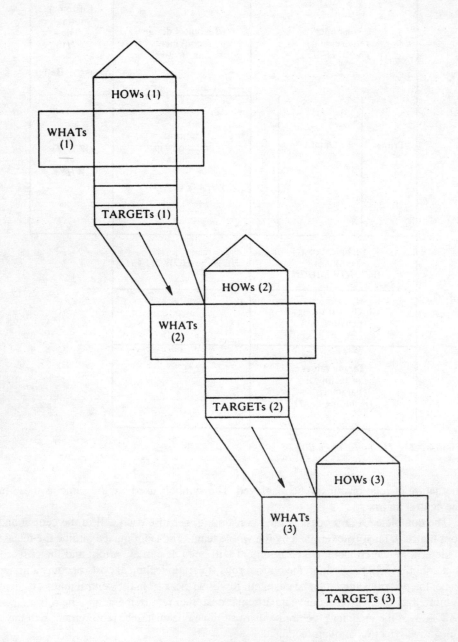

Figure 3.3 *The 'deployment' of the 'voice of the customer' through quality tables*

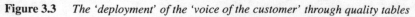

The very bottom of the house of quality diagram shows the *target* values of the *technical characteristics*, which are expressed in physical terms. They can only be decided by the team after discussion of the complete house contents. While these targets are the physical output of the QFD exercise, the whole process of information-gathering, structuring, and ranking generates a tremendous improvement in the team's cross-functional understanding of the product/service design delivery system. The target technical characteristics may be used to generate the next level house of quality diagram, where they become the WHATs, and the QFD process determines the further details of HOW they are to be achieved. In this way the process 'deploys' the customer requirements all the way to the final operational stages. Figure 3.3 shows how the target technical characteristics at each level becomes the input to the next level matrix.

QFD progresses now through the use of the 'seven new planning tools'[1] and other standard techniques such as value analysis,[2] experimental design,[3] statistical process control,[4] and so on.

The benefits of QFD

The aim of the HOQ is to co-ordinate the inter-functional activities and skills within an organisation. This should lead to products and services designed, produced/operated, and marketed so that customers will want to purchase them and continue doing so.

The use of competitive information in QFD should help to prioritise resources and to structure the existing experience and information. This allows the identification of items that can be acted upon.

There should be reductions in the number of midstream design changes, and these reductions in turn will limit post-introduction problems and reduce implementation time. Because QFD is consensus-based, it promotes teamwork and creates communications at functional interfaces, while also identifying required actions. It should lead to a 'global view' of the development process, from a consideration of all the details.

If QFD is introduced systematically, it should add structure to the information, generate a framework for sensitivity analysis, and provide documentation, which must be 'living' and adaptable to change. In order to understand the full impact of QFD it is necessary to examine the changes that take place in the team and the organisation during the design and development process. The main benefit of QFD is of course the increase in customer satisfaction, which may be measured in terms of, for example, reductions in warranty claims.

3.3 Design control

Design, like any other activity, must be carefully managed. A flowchart of the various stages and activities involved in the design and development process appears in Figure 3.4.

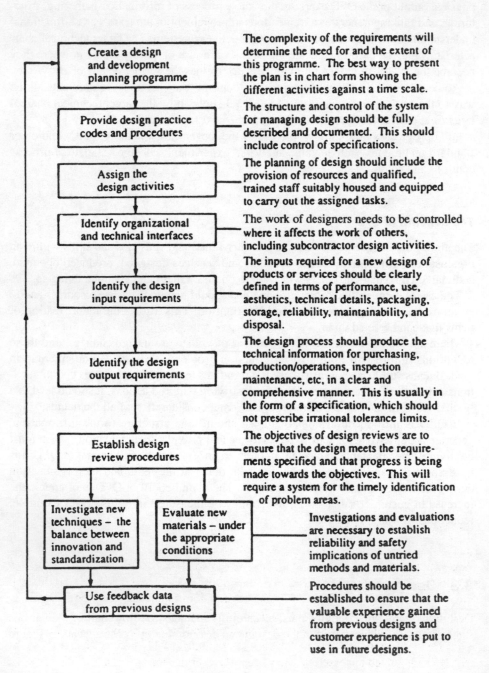

Create a design and development planning programme

The complexity of the requirements will determine the need for and the extent of this programme. The best way to present the plan is in chart form showing the different activities against a time scale.

Provide design practice codes and procedures

The structure and control of the system for managing design should be fully described and documented. This should include control of specifications.

Assign the design activities

The planning of design should include the provision of resources and qualified, trained staff suitably housed and equipped to carry out the assigned tasks.

Identify organizational and technical interfaces

The work of designers needs to be controlled where it affects the work of others, including subcontractor design activities.

Identify the design input requirements

The inputs required for a new design of products or services should be clearly defined in terms of performance, use, aesthetics, technical details, packaging, storage, reliability, maintainability, and disposal.

Identify the design output requirements

The design process should produce the technical information for purchasing, production/operations, inspection maintenance, etc, in a clear and comprehensive manner. This is usually in the form of a specification, which should not prescribe irrational tolerance limits.

Establish design review procedures

The objectives of design reviews are to ensure that the design meets the requirements specified and that progress is being made towards the objectives. This will require a system for the timely identification of problem areas.

Investigate new techniques – the balance between innovation and standardization

Evaluate new materials – under the appropriate conditions

Investigations and evaluations are necessary to establish reliability and safety implications of untried methods and materials.

Use feedback data from previous designs

Procedures should be established to ensure that the valuable experience gained from previous designs and customer experience is put to use in future designs.

Figure 3.4   *The design control process*

By structuring the design process in this way, it is possible to:

- Control the various stages.
- Check that they have been completed.
- Decide which management functions need to be brought in and at what stage.
- Estimate the level of resources needed.

The design control must be carefully handled to avoid stifling the creativity of the designer(s), which is crucial in making design solutions a reality.

It is clear that the design process requires a range of specialised skills, and the way in which these skills are managed, the way they interact, and the amount of effort devoted to the different stages of the design and development process is fundamental to the quality, producibility, and price of the service or final product. A QFD team approach to the management of design can play a major role in the success of a project.

It is never possible to exert the same tight control on the design effort as on other operational efforts, yet the cost and the time used are often substantial, and both must appear somewhere within the organisation's budget.

Certain features make control of design difficult:

1 No design will ever be 'complete' in the sense that, with effort, some modification or improvement cannot be made.
2 Few designs are entirely novel. An examination of most 'new' products, services or processes will show that they employ existing techniques, components or systems to which have been added a comparatively small novel element.
3 The longer the time spent on a design, the less the increase in the value of the design unless a technological breakthrough is achieved. This diminishing return from the design effort must be carefully managed.
4 External and/or internal customers will impose limitations on design time and cost. It is as difficult to imagine a design project whose completion date is not implicitly fixed, either by a promise to a customer, the opening of a trade show or exhibition, a seasonal 'deadline', a production schedule or, some other constraint, as it is to imagine an organisation whose funds are unlimited, or a product whose price has no ceiling.

Total design processes

Quality of design, then, concerns far more than the product or service design and its ability to meet the customer requirements. It is also about the activities of design and development. The appropriateness of the actual *design process* has a profound influence on the quality performance of any organisation, and much can be learned by examining successful organisations and how their strategies for research, design, and development are linked to the efforts of marketing and operations. In some quarters this is referred to as 'total design', and the term 'simultaneous engineering' has been used. This is an integrated approach to a new product or service introduction, similar in many ways to QFD in using multifunction teams or task forces to ensure that research, design, development, manufacturing, purchasing, supply, and marketing all work in parallel from concept through to the final launch of the product or service into the market place, including servicing and maintenance.

3.4   Specifications and standards

There is a strong relationship between standardisation and specification. To ensure that a product or a service is *standardised* and may be repeated a large number of times in exactly the manner required, *specifications* must be written so that they are open to only one interpretation. The requirements, and therefore the quality, must be built into the design specification. There are national and international standards which, if used, help to ensure that specifications will meet certain accepted criteria of technical or managerial performance, safety, etc.

Standardisation does not guarantee that the best design or specification is selected. It may be argued that the whole process of standardisation slows down the rate and direction of technological development, and affects what is produced. If standards are used correctly, however, the process of drawing up specifications should provide opportunities to learn more about particular innovations and to change the standards accordingly.

It is possible to strike a balance between innovation and standardisation. Clearly, it is desirable for designers to adhere where possible to past-proven materials and methods, in the interests of reliability, maintainability and variety control. Hindering designers from using recently developed materials, components, or techniques, however, can cause the design process to stagnate technologically. A balance must be achieved by analysis of materials, products and processes proposed in the design, against the background of their known reproducibility and reliability. If breakthrough innovations are proposed, then analysis or testing should be indicated objectively, justifying their adoption in preference to the established alternatives.

It is useful to define a specification. The International Standards Organisation (ISO) defines it in ISO 8402 (1986) as 'The document that prescribes the requirements with which the product or service has to conform'. A document not giving a detailed statement or description of the requirements to which the product, service or process must comply cannot be regarded as a specification, and this is true of much sales literature.

The specification conveys the customer requirements to the supplier to allow the product or service to be designed, engineered, produced, or operated by means of conventional or stipulated equipment, techniques, and technology. The basic requirements of a specification are that it gives the:

- Performance requirements of the product or service.
- Parameters – such as dimensions, concentration, turn-round time – which describe the product or service adequately (these should be quantified and include the units of measurement).
- Materials to be used by stipulating properties or referring to other specifications.
- Method of production or operations.
- Inspection/testing/checking requirements.
- References to other applicable specifications or documents.

To fulfil its purpose the specifications must be written in terminology that is readily understood, and in a manner that is unambiguous and so cannot be subject to differing interpretation. This is not an easy task, and one which requires all the expertise and knowledge available. Good specifications are usually the product of much discussion, deliberation and sifting of information and data, and represent tangible output from a QFD team.

3.5 Quality design in the service sector

The emergence of the services sector has been suggested by economists to be part of the natural progression in which economic dominance changes first from agriculture to manufacturing and then to services. It is argued that if income elasticity of demand is higher for services than it is for goods, then as incomes rise, resources will shift toward services. The continuing growth of services verifies this, and is further explained by changes in culture, fitness, safety, demography and life styles.

In considering the design of services it is important to consider the differences between goods and services. Some authors argue that the marketing and design of goods and services should conform to the same fundamental rules, whereas others claim that there is a need for a different approach to services because of the recognisable differences between the goods and services themselves.

In terms of design, it is possible to recognise three distinct elements in the service package – the physical elements or facilitating goods, the explicit service or sensual benefits, and implicit service or psychological benefits. In addition, the particular characteristics of service delivery systems may be itemised:

- Intangibility.
- Perishability.
- Simultaneity.
- Heterogeneity.

It is difficult, if not impossible, to design the intangible aspects of a service, since consumers often must use experience or the reputation of a service organisation and its representatives to judge quality.

Perishability is often an important issue in services, since it is often impossible or undesirable to hold stocks of the explicit service element of the service package. This aspect often requires that service operation and service delivery must exist simultaneously.

Simultaneity occurs because the consumer must be present before many services can take place. Hence, services are often formed in small and dispersed units, and it is difficult to take advantage of economies of scale. There is evidence that the emergence of computer and communications technologies is changing this in sectors such as banking, but contact continues to be necessary for the majority. Design considerations here include the environment and the systems used. Service facilities, procedures, and systems should be designed with the customer in mind, as well as the 'product' and the human resources. Managers need a picture of the total span of the operation, so that factors which are crucial to success are not neglected. This clearly means that the functions of marketing, design, and operations cannot be separated in services, and this must be taken into account in the design of the operational controls, such as the diagnosing of individual customer expectations. A QFD approach here is most appropriate.

Heterogeneity of services occurs in consequence of explicit and implicit service elements relying on individual preferences and perceptions. Differences exist in the outputs of organisations generating the same service, within the same organisation, and even the same employee on different occasions. Clearly, unnecessary variation needs to be controlled, but the variation attributed to estimating, and then matching, the consumers' requirements is essential to customer satisfaction and must be designed into the systems. This inherent variability does, however, make it difficult to set precise quantifiable standards for all the elements of the service.

In the design of services it is useful to classify them in some way. Several sources from the literature on the subject help us to place services in one of five categories:

- Service factory.
- Service shop.
- Mass service.
- Professional service.
- Personal services.

Several service attributes have particular significance for the design of service operations:

1 *Labour intensity* – the ratio of labour costs incurred to the value of plant and equipment used (people versus equipment-based services).
2 *Contact* – the proportion of the total time required to provide the service for which the consumer is present in the system.
3 *Interaction* – the extent to which the consumer actively intervenes in the service process to change the content of the service; this includes customer participation to provide information from which needs can be assessed, and customer feedback from which satisfaction levels can be inferred.
4 *Customisation* – which includes *choice* (providing one or more selections from a range of options, which can be single or fixed) and *adaptation* (the interaction process in which the requirement is decided, designed and delivered to match the need).
5 *Nature of service act* – either tangible, ie perceptible to touch and can be owned, or intangible, ie insubstantial.
6 *Recipient of service* – either people or things.

Table 3.1 gives a list of some services with their assigned attribute types and Table 3.2 shows how these may be used to group the services under the various classifications

Table 3.1 *A classification of selected services*

Service	Labour intensity	Contact	Inter-action	Custom-ization	Nature of act	Recipient of service
Accountant	High	Low	High	Adapt	Intangible	Things
Architect	High	Low	High	Adapt	Intangible	Things
Bank	Low	Low	Low	Fixed	Intangible	Things
Beautician	High	High	High	Adapt	Tangible	People
Bus service	Low	High	Low	Choice	Tangible	People
Cafeteria	Low	High	High	Choice	Tangible	People
Cleaning firm	High	Low	Low	Fixed	Tangible	Things
Clinic	Low	High	High	Adapt	Tangible	People
Coach service	Low	High	Low	Choice	Tangible	People
Sports coaching	High	High	High	Adapt	Intangible	People
College	High	High	Low	Fixed	Intangible	People
Courier firm	High	Low	Low	Adapt	Tangible	Things
Dental practice	High	High	High	Adapt	Tangible	People
Driving school	High	High	High	Adapt	Intangible	People
Equip. hire	Low	Low	Low	Choice	Tangible	Things
Finance consult.	High	Low	High	Adapt	Intangible	Things
Hairdresser	High	High	High	Adapt	Tangible	People
Hotel	High	High	Low	Choice	Tangible	People
Leisure centre	Low	High	High	Choice	Tangible	People
Maintenance	Low	Low	Low	Choice	Tangible	Things
Nursery	High	Low	Low	Fixed	Tangible	People
Optician	High	High	High	Adapt	Tangible	People
Postal service	Low	Low	Low	Adapt	Tangible	Things
Rail service	Low	High	Low	Choice	Tangible	People
Repair firm	Low	Low	Low	Adapt	Tangible	Things
Restaurant	High	High	Low	Choice	Tangible	People
Service station	Low	High	High	Choice	Tangible	People
Solicitors	High	Low	High	Adapt	Intangible	Things
Take away	High	Low	Low	Choice	Tangible	People
Veterinary	High	Low	High	Adapt	Tangible	Things

It is apparent that services are part of almost all organisations and not confined to the service sector. What is clear is that the service classifications and different attributes must be considered in any service design process.

(The authors are grateful to the contribution made by John Dotchin to this section of Chapter 3.)

Table 3.2   *Grouping of similar services*

PERSONAL SERVICES	

Driving school	Sports coaching
Beautician	Dental practice
Hairdresser	Optician

SERVICE SHOP

Clinic	Cafeteria
Leisure centre	Service station

PROFESSIONAL SERVICES

Accountant	Architect
Finance consultant	Solicitors
Veterinary	

MASS SERVICES

Hotel	Restaurant
College	Bus service
Coach service	Rail service
Take away	Nursery
Courier firm	

SERVICE FACTORY

Cleaning firm	Postal service
Repair firm	Equipment hire
Maintenance	Bank

Chapter highlights

Innovation, design and improvement

- All businesses need to update their products, processes and services.
- Innovation entails both invention and design, *and* continuous improvement of existing products, services, and processes.
- Leading product/service innovations are market-led. This requires a commitment at the top to building in quality throughout the design process. Moreover, the operational processes must be capable of achieving the design.
- State-of-the-art approach to innovation is based on a strategic balance of old and new, top management approach to design, and teamwork. The 'Styling' of products must also be matched by secondary design considerations, such as operating instructions and software support.

Quality function deployment (QFD) – the house of quality

- The 'house of quality' is the framework of the approach to design management known as quality function deployment (QFD). It provides structure to the design and development cycle, which is driven by customer needs rather than innovation in technology.

- QFD is a system for designing a product or service, based on customer demands, and bringing in all members of the supplier organisation.
- A QFD team's purpose is to take the needs of the market and translate them into such a form that they can be satisfied within the operating unit.
- The QFD team answers the following questions. WHO are the customers? WHAT do the customers need? HOW will the needs be satisfied?
- The answers to the WHO, WHAT and HOW questions are entered into the QFD matrix or quality table, one of the seven new tools of planning and design.
- The foundations of the house of quality are the customer requirements; the framework is the central planning matrix, which matches the 'voice of the customer' with the 'voice of the processes' (the technical descriptions and capabilities); and the roof is the interrelationships matrix between the technical design requirements.
- The benefits of QFD include customer-driven design, prioritising of resources, reductions in design changes and implementation time, and improvements in teamwork, communications, functional interfaces, and customer satisfaction.

Design control

- Design must be managed and controlled through planning, practice codes, procedures, activities assignments, identification of organisational and technical interfaces and design input requirements, review investigation and evaluation of new techniques and materials, and use of feedback data from previous designs.
- Total design or 'simultaneous engineering' is similar to QFD and uses multifunction teams to provide an integrated approach to product or service introduction.

Specifications and standards

- There is a strong relation between standardisation and specifications. If standards are used correctly, the process of drawing up specifications should provide opportunities to learn more about innovations and change standards accordingly.
- The aim of specifications should be to reflect the true requirements of the product/service that are capable of being achieved.

Quality design in the service sector

- In the design of services three distinct elements may be recognised in the service package: physical (facilitating goods), explicit service (sensual benefits), and implicit service (psychological benefits). Moreover, the characteristics of service delivery may be itemised as intangibility, perishability, simultaneity, and heterogeneity.
- Services may be classified generally as service factory, service shop, mass service, professional service, and personal service. The service attributes that are important in designing services include labour intensity, contact, interaction, customisation, nature of service act, and the direct recipient of the act.
- Use of this framework allows services to be grouped under the five classifications.

References

1   See J S Oakland, *Total Quality Management*, 2nd edition, Butterworth-Heinemann, 1993.
2   See D Samson, A Sohal, A Muhlemann, J Oakland and K Lockyer, *Production and Operations Management in Australia*, Longman, Melbourne, 1995.
3   See R Caulcutt, *Statistics in Research and Development*, 2nd edition, Chapman and Hall, 1991.
4   See J S Oakland and R F Followell, *Statistical Process Control*, 2nd edition, Butterworth-Heinemann, 1990.

Discussion questions

1 You are planning to open a wine bar, and have secured the necessary capital. Your aim is to attract both regular customers and passing trade. Discuss the key implications of this for the management of the business.

2 Discuss the following:
(a) the difference between quality and reliability
(b) total quality-related costs (Part 3)
(c) 100 per cent inspection

3 Discuss the various facets of the quality control function, paying particular attention to its interfaces with the other functional areas within the organisation.

4 Explain what you understand by the term 'Total Quality Management', paying particular attention to the following terms: quality, supplier/customer interfaces, process.

5 Present a 'model' for total quality management, describing briefly the various elements of the model.

6 You are an operations management consultant with particular expertise in the area of product design and development. You are at present working in projects for four firms:
(a) a chain of hotels
(b) a mail order goods firm
(c) a furniture manufacturer
(d) a road construction contractor
What factors do you consider are important generally in your area of specialisation? Compare and contrast how these factors apply to your four current projects.

7 Select one of the so-called 'gurus' of Quality Management, such as Juran, Deming, Crosby or Ishikawa, and explain his approach, with respect to the 'Oakland Model' of TQM. Discuss the strengths and weaknesses of the approach using this framework.

8 Discuss the application of the TQM concept in the service sector, paying particular attention to the nature of services and the customer–supplier interfaces.

9 In your new role as quality manager of the high-tech unit of a large national company, you identify a problem which is typified by the two internal memos shown below. Discuss in some detail the problems illustrated by this conflict, explaining how you would set about trying to make improvements:

From: Marketing Director
To: Managing Director
cc:
Production Director
Works Manager
Date: 6 August

We have recently carried out a customer survey to examine how well we are doing in the market. With regard to our product range, the reactions were generally good, but the 24-byte microwinkle thrystor is a problem. Without exception everyone we interviewed said that its quality is not good enough. Although it is not yet apparent, we will inevitably lose our market share.
As a matter of urgency, therefore, will you please authorise a complete redesign of this product?

From: Works Manager
To: Production Designer
Date: 6 August

This is really ridiculous!
I have all the QC records for the past 10 years on this product. How can there be anything seriously wrong with the quality when we only get 0.1 per cent rejects at final inspection and less than 0.01 per cent returns from customers?

10 Discuss the application of quality function deployment (QFD) and the 'house of quality' in:
    (a) a fast moving consumer goods (fmcg) company, such as one that designs, produces, and sells/distributes cosmetic products
    (b) an industrial company, such as one producing plastic materials
    (c) a commercial service organisation such as a bank or insurance company

Part Two

TQM – The Role of the Quality System

I must create a System or be enslaved by another man's.
William Blake, 1757–1827, from 'Jerusalem'

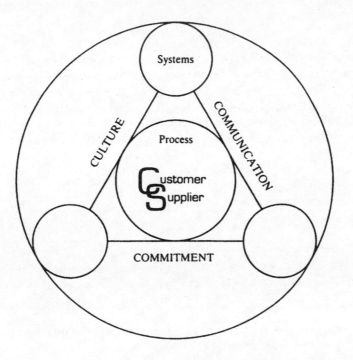

4
Planning for quality

4.1 Quality planning

Systematic planning is a basic requirement for effective quality management in all organisations. For quality planning to be useful, however, it must be part of a continuous review process that has as its objective zero errors or defectives, through a strategy of never-ending improvement. Before an appropriate total quality management system can be developed, it is necessary to carry out a preliminary analysis to ensure that a quality organisation structure exists, that the resources required will be made available, and that the various assignments will be carried out. This analysis has been outlined in the flowchart of Figure 4.1. The answers to the questions will generate the appropriate action plans.

In quality planning it is always necessary to review existing programmes within the organisation's functional areas, and these may be compared with the results of the preliminary analysis to appraise the strengths and weaknesses in quality throughout the business or operation. When this has been done, the required systems and programmes may be defined in terms of detailed operating plans, procedures and techniques. This may proceed through the flowchart of Figure 4.2, which provides a logical approach to developing a multifunctional total quality management system.

A quality plan

A quality plan is a document which is specific to each product, activity or service (or group) that sets out the necessary quality-related activities. The plan should include references to any:

- Purchased material or service specifications.
- Quality system procedures.
- Product formulation or service type.
- Process control.
- Sampling and inspection procedures.
- Packaging or distribution specifications.
- Miscellaneous, relevant procedures.

Such a quality plan might form part of a detailed operating procedure.

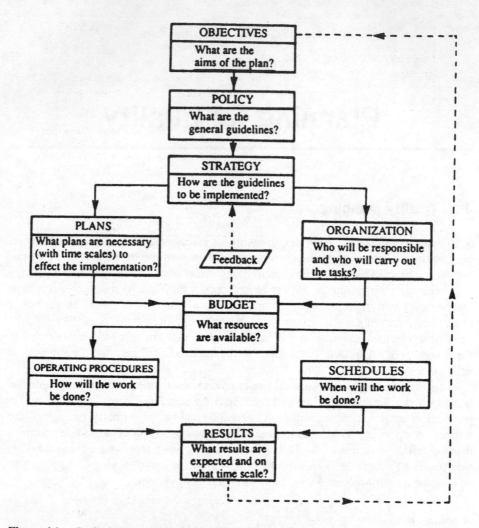

Figure 4.1   *Preliminary analysis for quality planning*

For projects relating to new products or services, or to new processes, written quality plans should be prepared to define:

1   Specific allocation of responsibility and authority during the different stages of the project.
2   Specific procedures, methods and instructions to be applied throughout the project.
3   Appropriate inspection, testing, checking, or audit programmes required at various defined stages.
4   Methods of changes or modifications in the plan as the project proceeds.

Some of the main points in the planning of quality relate very much to the *inputs* of processes:

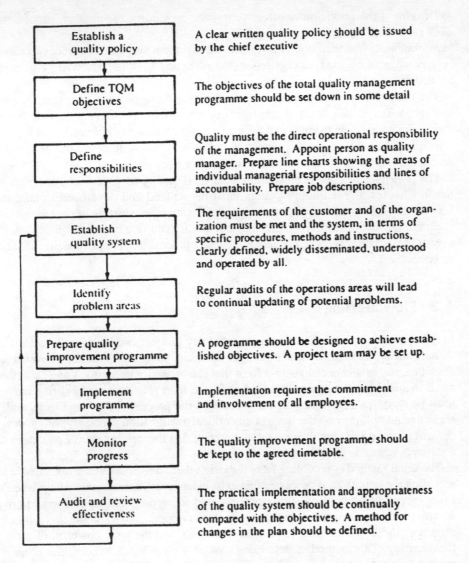

The flow chart (left boxes and their right-hand descriptions):

Establish a quality policy	A clear written quality policy should be issued by the chief executive
Define TQM objectives	The objectives of the total quality management programme should be set down in some detail
Define responsibilities	Quality must be the direct operational responsibility of the management. Appoint person as quality manager. Prepare line charts showing the areas of individual managerial responsibilities and lines of accountability. Prepare job descriptions.
Establish quality system	The requirements of the customer and of the organization must be met and the system, in terms of specific procedures, methods and instructions, clearly defined, widely disseminated, understood and operated by all.
Identify problem areas	Regular audits of the operations areas will lead to continual updating of potential problems.
Prepare quality improvement programme	A programme should be designed to achieve established objectives. A project team may be set up.
Implement programme	Implementation requires the commitment and involvement of all employees.
Monitor progress	The quality improvement programme should be kept to the agreed timetable.
Audit and review effectiveness	The practical implementation and appropriateness of the quality system should be continually compared with the objectives. A method for changes in the plan should be defined.

Figure 4.2   *Plan for a quality system*

Plant/equipment – the design, layout, and inspection of plant and equipment, including heating, lighting, storage, disposal of waste, etc.

Processes – the design and monitoring of processes to reduce to a minimum the possibility of malfunction and/or failure.

Workplace – the establishment and maintenance of suitable, clean and orderly places of work.

Facilities – the provision and maintenance of adequate facilities.

Procedures – the preparation of procedures for all operations. These may be in the form of general plans and guides rather than tremendous detail, but they should include specific operational duties and responsibilities.

Training – the provision of effective training in quality, technology, process and plant operation.

Information – the lifeblood of all quality management systems. All processes should be accompanied by good data collection, recording and analysis, followed by appropriate action.

The quality plan should focus on providing action to prevent cash leaking away through waste. If the quality management system does not achieve this, then there is something wrong with the plan and the way it has been set up or operated – not with the principle. The whole approach should be methodical and systematic, and designed to function irrespective of changes in management or personnel.

The principles and practice of setting up a good quality-management system are set out in Chapter 5. The quality system must be planned and developed to take into account all other functions, such as design, development, production or operations, sub-contracting, installation, maintenance, and so on. The remainder of this chapter is devoted to certain aspects of the quality-planning process that require specific attention or techniques.

4.2 Flowcharting

In the systematic planning or examination of any process, whether that be a clerical, manufacturing, or managerial activity, it is necessary to record the series of events and activities, stages and decisions in a form that can be easily understood and communicated to all. If improvements are to be made, the facts relating to the existing method must be recorded first. The statements defining the process should lead to its understanding and will provide the basis of any critical examination necessary for the development of improvements. It is essential therefore that the descriptions of processes are accurate, clear and concise.

The usual method of recording facts is to write them down, but this is not suitable for recording the complicated processes that exist in any organisation, particularly when an exact record is required of a long process, and its written description would cover several pages requiring careful study to elicit every detail. To overcome this difficulty, certain methods of recording have been developed, and the most powerful of these is flowcharting. This method of describing a process owes much to computer programming, where the technique is used to arrange the sequence of steps required for the operation of the programme. It has a much wider application, however, than computing.

Certain standard symbols are used on the chart, and these are shown in Figure 4.3. The starting point of the process is indicated by a circle. Each processing step, indicated by a rectangle, contains a description of the relevant operation, and where the process ends is indicated by an oval. A point where the process branches because of a decision is shown by a diamond. A parallelogram contains useful information but is not a processing step. The arrowed lines are used to connect symbols and to indicate direction of flow. For a complete description of the process all operation steps (rectangles) and decisions (diamonds) should be connected by pathways to the start circle and end oval. If the flowchart cannot be drawn in this way, the process is not fully understood.

It is a salutary experience for most people to sit down and try to draw the flowchart for a process in which they take part every working day. It is often found that:

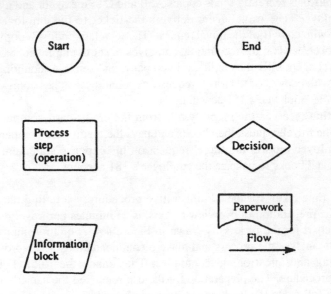

Figure 4.3 *Flowcharting symbols*

- The process flow is not fully understood.
- A single person is unable to complete the flowchart without help from others.

The very act of flowcharting will improve knowledge of the process, and will begin to develop the teamwork necessary to find improvements. In many cases the convoluted flow and octopus-like appearance of the chart will highlight unnecessary movements of people and materials and lead to commonsense suggestions for waste elimination.

Example of flowcharting in use – Improving a travel procedure

We start by describing the original process for a male employee, though clearly it applies equally to females.

The process starts with the employee explaining his travel plans to his secretary. The secretary then calls the travel agent to inquire about the possibilities and gives feedback to the employee. The employee decides if the travel arrangements, eg flight numbers and dates, are acceptable and informs his secretary, who calls the agent to make the necessary bookings or examine alternatives. The administrative procedure, which starts as soon as the bookings have been made, is as follows:

1 The employee's secretary prepares the travel request (which is in four parts, A, B, C and D), and gives it to his secretary. The request is then sent to the employee's manager, who approves it. The manager's secretary sends it back to the employee's secretary.

2   The employee's secretary sends copies A, B and C to the agent and gives copy D to the employee. The travel agent delivers the ticket to the employee's secretary, together with copy B of the travel request. The secretary endorses copy B for receipt of the ticket, sends it to Accounting, and gives ticket to employee.
3   The travel agent bills the credit-card company, and sends Accounting a pro-forma invoice with copy C of the travel request. Accounting matches copies B and C, and charges the employee's 181 account.
4   Accounting receives the monthly bill from the credit-card company, matches it against the travel request, then books and pays the credit-card company.
5   The employee reports the travel request on his expense statement. Accounting matches and books to balance the employee's 181 account.

The total time taken for the administrative procedure, excluding the correction of errors and the preparation of overview reports, is 23 minutes per travel request.

The flowchart for the process is drawn in Figure 4.4. A quality-improvement team was set up to analyse the process and make recommendations for improvement, using brainstorming and questioning techniques. They made the following proposal to change the procedure. The preparation for the trip remained the same but the administrative steps, following the bookings being made became:

1   The travel agent sends the ticket to the secretary, along with a receipt document, which is returned to the agent with the secretary's signature.
2   The agent sends the receipt to the credit-card company, which bills the company on a monthly basis with a copy of all the receipts. Accounting pays the credit-card company and charges the employee's 181 account.
3   The employee reports the travel on his expense statement, and Accounting books to balance the employee's 181 account.

The flowchart for the improved process is shown in Figure 4.5. The proposal reduced the total administrative effort per travel request (or per travel arrangement, because the travel request was eliminated) from 23 minutes to 5 minutes.

The details that appear on a flowchart for an existing process must be obtained from direct observation of the process, not by imagining what is done or what should be done. The latter may be useful, however, in the planning phase, or for outlining the stages in the introduction of a new concept. Such an application is illustrated in Figure 4.6 for the installation of statistical process control charting systems (see Chapter 8). Similar charts may be used in the planning of quality management systems.

It is surprisingly difficult to draw flowcharts for even the simplest processes, particularly managerial ones, and following the first attempt it is useful to ask whether:

*   The facts have been correctly recorded.
*   Any over-simplifying assumptions have been made.
*   All the factors concerning the process have been recorded.

The authors have seen too many process flowcharts that are so incomplete as to be grossly inaccurate.

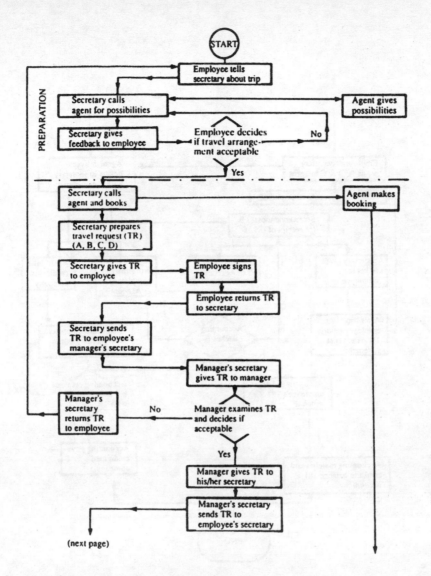

Figure 4.4 *Original process for travel procedure*

Summarising, then, a flowchart is a picture of the steps used in performing a function. This function can be anything from a process step to accounting procedures, even preparing a meal. Lines connect the steps to show the flow of the various functions. Flowcharts provide excellent documentation and are useful trouble-shooting tools to determine how each step is related to the others. By reviewing the flowchart, it is often possible to discover inconsistencies and determine potential sources of variation and problems. For this reason, flowcharts are very useful in process improvement when examining an existing process to highlight the problem areas. A group of people, with the knowledge about the process, should take the following simple steps:

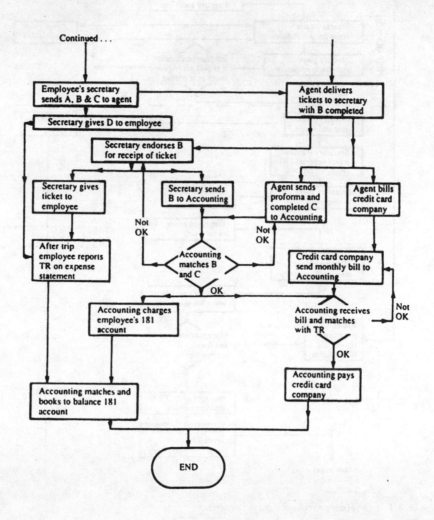

Figure 4.4   *(continued)*

1   Draw a flowchart of existing process.
2   Draw a second chart of the flow the process could or should follow.
3   Compare the two to highlight the changes necessary.

4.3 Planning for purchasing

Very few organisations are self-contained to the extent that their products and services are all generated at one location, from basic materials. Some materials or services are usually purchased from outside organisations, and the primary objective of purchasing is to obtain the correct equipment, materials, and services in the right quantity, of the right quality, from the right origin, at the right time and cost. Purchasing also plays a vital role as the organisation's 'window-on-the-world', providing information on any new products, processes, materials and services. It should also advise on probable

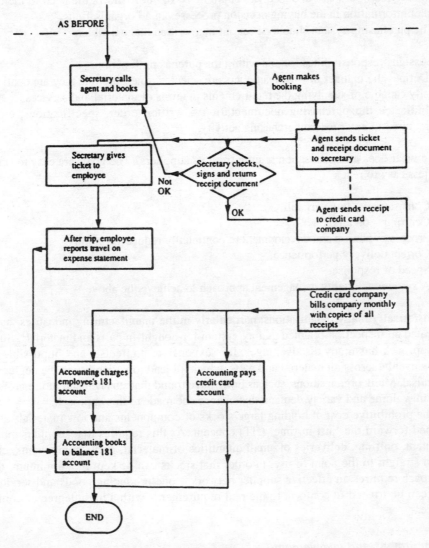

Figure 4.5   *Improved process for travel procedure*

prices, deliveries, and performance of products under consideration by the research, design and development functions.

Although purchasing is clearly an important area of managerial activity, it is often neglected by both manufacturing and service industries. The separation of purchasing from selling has, however, been removed in many large retail organisations, which have recognised that the purchaser must be responsible for the whole 'product line' – its selection, quality, specification, delivery, price, acceptability, and reliability. If any part of this chain is wrong, the purchasing function must resolve the problem. This concept is clearly very appropriate in retailing, where transformation activities on the product itself, between purchase and sale, are small or zero, but it shows the need to include market information in the buying decision processes in all organisations.

The purchasing system should be set out in a written manual which:

1   Assigns responsibilities for and within the purchasing function.
2   Defines the manner in which suppliers are selected, to ensure that they are continually capable of supplying the requirements in terms of material and services.
3   Indicates the purchasing documentation – written order, specifications, etc – required in any modern purchasing activity.

So what does an organisation require from its suppliers? The goals are easy to state, but less easy to reach:

• Consistency – low variability.
• Centring – on target.
• Process evolution and development to continually reduce variability.
• Correct delivery performance.
• Speed of response.
• A *systematic* quality management approach to achieve the above.

Historically many organisations, particularly in the manufacturing industries, have operated an inspection-oriented quality system for bought-in parts and materials. Such an approach has many disadvantages. It is expensive, imprecise, and impossible to apply evenly across all material and parts, which all lead to variability in the degree of appraisal. Many organisations, such as Ford, have found that survival and future growth in both volume and variety demand that changes be made to this approach.

The prohibitive cost of holding large stocks of components and raw materials also pushed forward the 'just-in-time' (JIT) concept. As this requires that suppliers make frequent, on time, deliveries of small quantities of material, parts, components, etc, often straight to the point of use, in order that stocks can be kept to a minimum, the approach requires an effective supplier network – one producing goods and services that can be trusted to conform to the real requirements with a high degree of confidence.

Commitment and involvement

The process of improving suppliers' performance is complex and clearly relies very heavily on securing real commitment from the senior management of the supplier

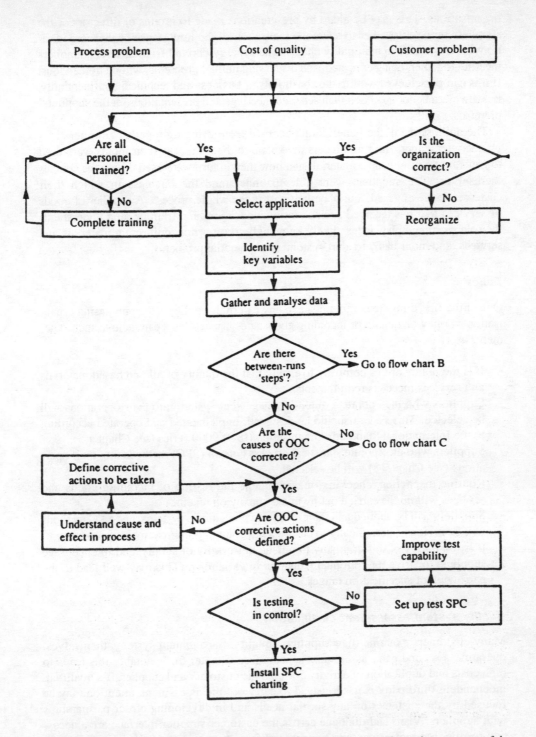

Figure 4.6 *Flowchart (A) for installation of SPC charting systems. (The authors are grateful to Exxon Chemical International for permission to use and modify this chart.)*

organisations. This may be aided by presentations made to groups of directors of the suppliers brought together to share the realisation of the importance of their organisations' performance in the quality chains. The synergy derived from different suppliers meeting together, being educated, and discussing mutual problems, will be tremendous. If this can be achieved, within the constraints of business and technical confidentiality, it is always a better approach than separate meetings and presentations on the suppliers' premises.

The authors recall the benefits that accrued from bringing together suppliers of a photocopier, paper, and ring binders to explain to them the way their inputs were used to generate training-course materials and how they in turn were used during the courses themselves. The suppliers were able to understand the business in which their customers were engaged, and play their part in the whole process. A supplier of goods *or* services that has received such attention, education, and training, and understands the role its inputs play, is less likely knowingly to offer nonconforming materials and services, and more likely to alert customers to potential problems.

Policy

One of the first things to communicate to any external supplier is the purchasing organisation's policy on quality of incoming goods and services. This can include such statements as:

- It is the policy of this company to ensure that the quality of all purchased materials and services meets its requirements.
- Suppliers who incorporate a quality management system into their operations will be selected. This system should be designed, implemented and operated according to the International Standards Organisation (ISO) 9000 series (see Chapter 5).
- Suppliers who incorporate statistical process control (SPC) methods into their operations (see Chapter 8) will be selected.
- Routine inspection, checking, measurement and testing of incoming goods and services will *not* be carried out by this company on receipt.
- Suppliers will be audited and their operating procedures, systems, and SPC methods will be reviewed periodically to ensure a never-ending improvement approach.
- It is the policy of this company to pursue uniformity of supply, and to encourage suppliers to strive for continual reduction in variability. (This may well lead to the narrowing of specification ranges.)

Quality system assessment certification

Many customers examine their suppliers' quality management systems themselves, operating a second party assessment scheme (see Chapter 6). Inevitably this leads to high costs and duplication of activity, for both the customer and supplier. If a qualified, independent third party is used instead to carry out the assessment, attention may be focused by the customer on any special needs and in developing closer partnerships with suppliers. Visits and dialogue across the customer-supplier interface are a necessity for the true requirements to be met, and for future growth of the whole business chain. Visits should be concentrated, however, on improving understanding and capability, rather than on close scrutiny of operating procedures, which is best left to experts, including those within the supplier organisations charged with carrying out internal system audits and reviews.

4.4 Planning for just-in-time (JIT) management

There are so many organisations throughout the world that are looking at, introducing, or practising just-in-time (JIT) management principles that the probability of encountering it is very high. JIT, like many modern management concepts, is credited to the Japanese, who developed and began to use it in the late 1950s. It took approximately 20 years for JIT methods to reach Western hardgoods industries and a further 10 years before businesses realised the generality of the concepts.

Basically JIT is a programme directed towards ensuring that the right quantities are purchased or produced at the right time, and that there is no waste. Anyone who perceives it purely as a material-control system, however, is bound to fail with JIT. JIT fits well under the TQM umbrella, for many of the ideas and techniques are very similar and, moreover, JIT will not work without TQM in operation. Writing down a definition of JIT for all types of organisation is extremely difficult, because the range of products, services and organisation structures leads to different impressions of the nature and scope of JIT. It is essentially:

- A series of operating concepts that allows systematic identification of operational problems.
- A series of technology-based tools for correcting problems following their identification.

The Kanban system

Kanban is a Japanese word meaning visible record, but in the West it is generally taken to mean a card that signals the need to deliver or produce more parts or components. In manufacturing, various types of record cards, eg job orders or tickets and route cards, are used for ordering more parts in a *push* type, schedule-based system. In a push system a multi-period master production schedule of future demands is prepared, and a computer explodes this into detailed schedules for producing or purchasing the appropriate parts or materials. The schedules then *push* the production of the parts or components, out and onward. These systems, when computer-based, are usually called Material Requirements Planning (MRP) or the more recent Manufacturing Resource Planning (MRPII).

The main feature of the Kanban system is that it *pulls* parts and components through the production processes when they are needed. Each material, component, or part has its own special container designed to hold a precise, preferably small, quantity. The number of containers for each part is a carefully considered management decision. Only standard containers are used, and they are always filled with the prescribed quantity.

A Kanban system provides parts when they are needed but without guesswork, and therefore without the excess inventory that results from bad guesses. The system will only work well, however, within the context of a JIT system in general, and the reduction of set-up times and lot sizes in particular. A JIT programme can succeed without a Kanban-based operation, but Kanbans will not function effectively independently of JIT.

Just-in-time purchasing

Purchasing is an important feature of JIT. The development of long-term relationships with a few suppliers, rather than short-term ones with many, leads to the concept of *co-producers* in networks of trust providing dependable quality and delivery of goods and services. Each organisation in the chain of supply is encouraged to extend JIT methods to its suppliers. The requirements of JIT mean that suppliers are usually located near the purchaser's premises, delivering small quantities, often several times per day, to match the usage rate. Paperwork is kept to a minimum and standard quantities in standard containers are usual. The requirement for suppliers to be located near the buying organisation, which places those at some distance at a competitive disadvantage, causes lead times to be shorter and deliveries to be more reliable.

It can be argued that JIT purchasing and delivery are suitable mainly for assembly line operations, and less so for certain process and service industries, but the reduction in the inventory and transport costs that it brings should encourage innovations to lead to its widespread adoption. Those committed to open competition and finding the lowest price will find most difficulty. Nevertheless, there must be a recognition of the need to develop closer relationships and to begin the dialogue – the sharing of information and problems – that leads to the product or service of the right quality, being delivered in the right quantity, at the right time.

Chapter highlights

Quality planning

- Systematic planning is a basic requirement for TQM.
- A quality plan sets out details for systems, procedures, purchased materials or services, products/services, plant/equipment, process control, sampling/inspection, training, packaging and distribution.

Flowcharting

- Flowcharting is a method of describing a process in pictures, using symbols – rectangles for operation steps, diamonds for decision, parallelograms for information, and circles/ovals for the start/end points. Arrow lines connect the symbols to show the 'flow'.
- Flowcharting improves knowledge of the process and helps to develop the team of people involved.
- Flowcharts document processes and are useful as trouble-shooting tools and in process improvement. An improvement team would flowchart the existing process

and the improved or desired process, comparing the two to highlight the changes necessary.

Planning for purchasing

- The prime objective of purchasing is to obtain the correct equipment, materials, and services in the right quantity, of the right quality, from the right origin, at the right time and cost. Purchasing also acts as a 'window-on-the-world'.
- The separation of purchasing from selling has been eliminated in many retail organisations, to give responsibility for a whole 'product line'. Market information must be included in *any* buying decision.
- The purchasing system should be set out in a written manual, which gives responsibilities, the means of selecting suppliers, and the documentation to be used.
- An organisation requires from its suppliers consistency, on target, process evolution, good delivery performance, speed of response, and systematic quality management.
- Improving supplier performance requires from the suppliers' senior management commitment, education, a policy, an assessed quality system, and supplier approval.

Planning for just-in-time (JIT) management

- JIT fits well under the TQM umbrella and is essentially a series of operating concepts that allow the systematic identification of problems, *and* tools for correcting them.
- Kanban cards signal the need to deliver or produce more parts or components. The system of Kanbans will work well only in the context of JIT.
- Purchasing is an important feature of JIT. Long-term relationships with a few suppliers, or 'co-producers', are developed in networks of trust to provide quality goods and services.

5

System design and contents

5.1 Why a documented system?

In earlier chapters we have seen how the keystone of quality management is the concept of customer and supplier working together for their mutual advantage. For any particular organisation this becomes 'total' quality management if the supplier/customer interfaces extend beyond the immediate customers, back inside the organisation, and beyond the immediate suppliers. In order to achieve this, a company must organise itself in such a way that the human, administrative and technical factors affecting quality will be under control. This leads to the requirement for the development and implementation of a quality system that enables the objectives set out in the quality policy to be accomplished. Clearly, for maximum effectiveness and to meet individual customer requirements, the quality system in use must be appropriate to the type of activity and product or service being offered.

It may be useful to reflect on why such a device is necessary to achieve control of processes. One of the authors remembers being at a table in a restaurant with eight people who all ordered the 'Chef's Special Individual Soufflé'. All eight soufflés arrived together at the table, magnificent in their appearance and consistency, each one exhibiting an almost identical size and shape – a truly remarkable demonstration of culinary skill. How had this been achieved? The chef had *managed* such consistency by making sure that, for each soufflé, he used the same ingredients (materials), the same equipment (plant), the same method (procedure) in exactly the same way every time. The process was under control. This is the aim of a good quality system, to provide the 'operator' of the process with consistency and satisfaction in terms of methods, materials, equipment, etc (Figure 5.1). Two feedback loops are also required: the 'voice' of the customer (marketing activities) and the 'voice' of the process (measurement activities).

The chef's soufflés were not ISO Standard, Australian Standard, or British Standard soufflés – they were the chef's special soufflés. It is not conceivable that the chef sat down with a blank piece of paper to invent a soufflé recipe. Why reinvent wheels? He probably used a standard formula and changed it slightly to make it his own. This is exactly the way in which organisations must use the international standards on quality systems that are available. The 'wheel' has been invented but it must be built in a way that meets the specific organisational and product or service requirements. The International Organisation for Standardisation (ISO) Standard 9000 Series (1994) sets out the methods by which a management system, incorporating all the activities associated with quality, can be implemented in an organisation to ensure that all the specified performance requirements and needs of the customer are fully met.

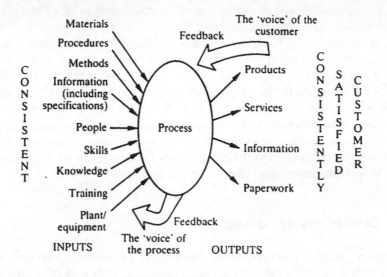

Figure 5.1   *The systematic approach to process management*

Let us return to the chef in the restaurant and propose that his success leads to a desire to open eight restaurants in which are served his special soufflés. Clearly he cannot rush from each one of these establishments to another every evening making soufflés. The only course open to him to ensure consistency of output, in all eight restaurants, is for him to write down in some detail the system he uses, and then make sure that it is used on all sites, every time a soufflé is produced. Moreover, he must periodically visit the different sites to ensure that:

1 The people involved are operating according to the documented system (a system audit).
2 The soufflé system still meets the requirements (a system review).

If in his system audits and reviews he discovers that an even better product or less waste can be achieved by changing the method or one of the materials, then he may wish to effect a change. To maintain consistency, he must ensure that the appropriate changes are made to the documented system, *and* that everyone concerned is issued with the revision and begins to operate accordingly.

A fully documented quality management system will ensure that two important requirements are met:

- *The customer's requirements* – for confidence in the ability of the organisation to deliver the desired product or service consistently.
- *The organisation's requirements* – both internally and externally, and at an optimum cost, with efficient utilisation of the resources available – material, human, technological, and administrative.

These requirements can be truly met only if objective evidence is provided, in the form of information and data, which supports the system activities, from the ultimate supplier through to the ultimate customer.

A *quality system* may be defined, then, as an assembly of components, such as the organisational structure, responsibilities, procedures, processes and resources for implementing total quality management. These components interact and are affected by being in the system, so the isolation and study of each one in detail will not necessarily lead to an understanding of the system as a whole. Often the interactions between the components – such as materials and processes, procedures and responsibilities – are just as important as the components themselves, and problems can arise from these interactions as much as from the components. Clearly, if one of the components is removed from the system, the whole thing will change.

5.2 Quality system design

The quality system should apply to and interact with all activities of the organisation. It begins with the identification of the requirements and ends with their satisfaction, at every transaction interface. The activities may be classified in several ways – generally as processing, communicating and controlling, but more usefully and specifically as:

1 Marketing.
2 Market research.
3 Design.
4 Specifying.
5 Development.
6 Procurement.
7 Process planning.
8 Process development and assessment.
9 Process operation and control.
10 Product- or service-testing or checking.
11 Packaging (if required).
12 Storage (if required).
13 Sales.
14 Distribution or installation/operation.
15 Technical service.
16 Maintenance.

These may be regarded as slats on a rotating drum rolling towards a satisfied customer, who becomes 'delighted' by the consistency of the product or service (Figure 5.2). The driving force of the drum is the centralised quality system, and the drum will not operate until the system is in place and working. The first step in getting the drum rolling is to prepare the necessary documentation. This means, in very basic terms, that procedures should be written down, preferably in such a way that the system conforms to one of the national or international standards. This is probably best done in the form of a quality manual.

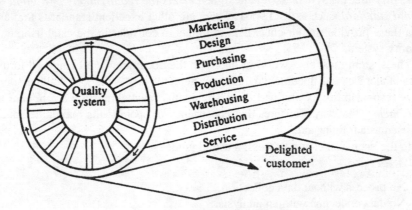

Figure 5.2   *The quality system power unit to the delighted 'customer'*

It is interesting to bring together the concept of Deming's Cycle of continuous improvement – PLAN DO CHECK ACT – and quality systems. A simplification of what a good quality system is trying to do is given in Figure 5.3, which follows the improvement cycle.

In most organisations established methods of working already exist, and all that is required is the writing down of what is currently done. In some instances companies may not have procedures to satisfy the requirements of a good standard, and they may have to begin to devise them. Alternatively, it may be found that two people, supposedly performing the same task, are working in different ways, and there is a need to standardise the procedure. Some organisations use the effective slogan 'If it isn't written down, it doesn't exist'. This can be a useful discipline, provided it doesn't lead to paper bureaucracy.

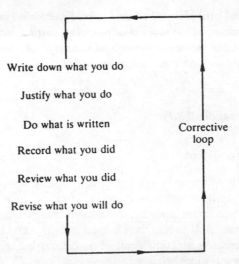

Figure 5.3   *The quality system and never ending improvement*

Justify that the *system* as it is designed *meets the requirements of a good international standard*, such as ISO 9001. There are other excellent standards that are used, and these provide similar checklists of things to consider in the establishment of the quality system.

The system must be a working one with the documents well fingered in use. One person alone cannot document a quality system; the task is the job of all personnel who have responsibility for any part of it. The quality manual must be a *practical working document* – that way it ensures that consistency of operation is maintained and it may be used as a training aid.

In the operation of any process, a useful guide is:

- No process without data collection.
- No data collection without analysis.
- No analysis without decisions.
- No decisions without actions, which can include doing nothing.

This excellent discipline is built into any good quality system, primarily through the audit and review systems. The requirement to *audit or 'check'* that the system is functioning according to plan, and to *review* possible system improvements, utilising audit results, should ensure that the *improvement* cycle is engaged through the *corrective action* procedures. The overriding requirement is that the systems must reflect the established practices of the organisation, improved where necessary to bring them into line with current and future requirements.

5.3   Quality system requirements

The special methods and procedures that need to be documented and implemented will be determined by the nature of the process or processes carried out. Certain fundamental principles are applicable, however, throughout industry, commerce, and the services. These fall into generally well defined categories, as follows.

1   Management responsibility

Quality policy (see Chapter 2)

The organisation should define and publish its quality policy, which forms one element of the corporate policy. Full commitment is required from the most senior management to ensure that the policy is communicated, understood, implemented and maintained at all levels in the organisation. It should therefore be authorised by top management and signed by the Chief Executive, or equivalent, who must also ensure that it is updated as appropriate to meet organisational changes.

Organisation

Organisations should have an organisation chart, and define the responsibilities of those shown in the chart, which should include all functions that affect quality. One manager with the necessary authority, resources, support, and ability should be given the responsibility to co-ordinate, implement, and monitor the quality system, resolve any problems and ensure prompt and effective corrective action. This includes responsibility for ensuring proper handling of the quality system. Those who control sales, service, processing, warehousing, delivery, and reworking of nonconforming product or service must also be identified.

Management review

Management reviews of the system must be carried out, with records to indicate the actions decided upon. The effectiveness of these actions should be considered during subsequent reviews. Reviews typically include data on the internal quality audits, customer complaints, non-conforming materials, the performance of sub-contractors, and the training plan.

Table 5.1   *Quality management and quality assurance standards ISO 9000 series*

ISO Standard	Title
	Principal concepts and applications
9000-1	Guidelines for selection and use.
9000-2	Generic guidelines for application of ISO 9001, 9002, 9003. Quality systems specifications:
9001	for design, development, production, installation and servicing,
9002	for production, installation and servicing,
9003	for final inspection and test.
	Guide to the use of BS5750: Part 1 'Specification for design/development, production, installation and servicing', Part 2 'Specification for production and installation', Part 3 'Specification for final inspection and test'.
9004	Guide to quality management and quality system elements.
9004-1	Guidelines.
9004-2	Guide to quality management and quality system elements for services.
9004-3	Guide for processed materials.
9004-4	Guide to quality improvement.
9004-5	Guide to use of quality plans.
9004-6	Guide to configuration maagement.
9000-3	Guide to the application of BS5750: Part 1 to the development, supply, and maintenance of software.
9000-4	Guide to dependability programme management.

2   *Quality system*

The organisation should prepare a quality plan and a quality manual that is appropriate for the 'level' of quality system required.

A *Level 1* system relates to design, production or operation, and installation, and applies when the customer specifies the goods or services in terms of how they are to perform, rather than in established technical terms.

A *Level 2* system is relevant when an organisation is producing goods or services to a customer's or a published specification.

A *Level 3* system applies only to final production or service inspection, check, or test procedures.

The reader is referred to the International Standard, ISO 9000 series and Table 5.1.

A quality manual should set out the general quality policies, procedures, and practices of the organisation. In the quality manual for large organisations it may be convenient to indicate simply the existence and contents of other manuals, those containing the details of procedures and practices in operation in specific areas of the system.

Before an organisation can agree to supply to a specification, it must ensure that:

(a) The processes and equipment (including any that are subcontracted) are capable of meeting the requirements.
(b) The operators have the necessary skills and training.
(c) The operating procedures are written down and not simply passed on verbally.
(d) The plant and equipment instrumentation is capable (eg measuring the process variables with the appropriate accuracy and precision).
(e) The quality-control procedures and any inspection, check, or test methods available, provide results to the required accuracy and precision, and are documented.
(f) Any subjective phrases in the specification, such as 'finely ground', 'low moisture content', 'in good time', are understood, and procedures to establish the exact customer requirements exist.

3   *Contract review*

It is difficult to over-emphasise the importance of this aspect of the system. Each accepted customer order should be regarded as a contract, and 'order entry procedures' should be developed and documented. These should ensure that:

(a) The customer requirements are absolutely clear and in writing, including the recording of any verbal communication, eg telephone instructions or orders.
(b) Differences between the order and any original inquiry and/or quotation are agreed or resolved.
(c) The terms of the order (contract) can be met, including verifying that dates promised to customers on acceptance of the contract can be met.

Clearly, a procedural dialogue should be established between customer and supplier with regard to the specification, interfaces and the communication of changes. The system must ensure that everyone in the organisation understands the commitment, skills, and resources required to meet any particular contract, and that these have been scheduled.

4   Design control

Where a Level 1 system is required, there must be procedures that control and verify design of products or services to ensure that the customer requirements will be met. The translation of the information derived from market research into practical designs that are achievable by the operating units should be the core of the documented design-control system. This will include the following activities, which were dealt with in more detail in Chapter 3:

- Planning of research, design and development.
- Assignment of design activities to qualified staff.
- Identification of the organisational and technical interfaces between different groups.
- Preparation of a design brief relating to the requirements of the product or service (inputs).
- Production of clear and comprehensive technical data to enable complete operation of the service or production and delivery of the product, according to the requirements (outputs).
- Verification that the design outputs meet the requirements of the inputs.
- Identification and control of all design changes and modifications and the associated documentation.

Attention, in detail, to the above areas will form the basis of a research, design and development programme. With correct implementation it will maintain a balance between innovation and standardisation, which encourages the use of new techniques whilst retaining reliable, proven designs, materials and methods.

5   Document and data control

All documents relating to quality, including the following, should be 'controlled':

(a) Quality manual and supplementary manuals.
(b) Departmental operating manuals.
(c) Written procedures.
(d) Purchasing specifications.
(e) Lists of approved suppliers.
(f) Product, parts, or service formulations.
(g) Intermediate, part, or component formulations and specifications.
(h) Service manuals.
(i) Relevant international and national standards.

'Control' is necessary to ensure that only the most up-to-date issues are used and referred to at the various locations. Clearly this will require records of what documents exist and/or are needed, and of who holds the documents; plus a written procedure for the issue of amendments and revisions, and for re-issues, together with some form of acknowledgment of receipt. Computerised techniques may be very helpful here. If additional copies of 'controlled' documents are produced for temporary purposes, procedures should exist to prevent their misuse. Sales literature is not usually regarded as controlled documentation, unless it forms the basis of a contract. In industries where continuous innovation, redesign and/or improvements are major features, good document change control is vital.

6   *Purchasing (see also Chapter 4)*

The objective of the purchasing system is very simple – to ensure that purchased products and services conform to the requirements of the organisation. The means of achieving this should be concentrated on assessments of the suppliers' own quality systems, rather than by an elaborate scheme of checking, testing and inspection on receipt. The system should essentially consider the 'contract review' from the view of the purchaser.

Suppliers or subcontractors should be selected on the basis of their ability to meet the defined requirements, and objective documentary evidence will be required to show that the supplier:

(a) Has the capability to do so.
(b) Will do so reliably and consistently.

When the extent of *vendor appraisal* necessary has to be decided, the following factors should be taken into consideration:

• Feasibility of appraisal.
• Objective evidence, from records and analysis of acceptable past performance (not 'reputation', which is subjective).
• Any independent third-party quality system assessment, certification, or registration to a recognised standard, eg ISO 9000 series.
• Any assessments by means of questionnaires or visits, which should be documented.

Vendor appraisal of any product or service subcontractors is often necessary to ensure that their quality system matches the standard of the purchasing organisation, and appraisal visits may form part of the corrective action following unsatisfactory performance. The basic requirements of purchasing documents are that they:

(a) Are written.
(b) Include the specification, or reference to it, to describe clearly the product/service required.
(c) Are made available to the supplier.
(d) Are reviewed and approved by authorised personnel before issue.

Purchases can be made by telephone or computer means, but must refer to the appropriate purchasing documentation. The system should allow customers to impose quality-system requirements on their suppliers' suppliers, and so on, if specified in the contract. This may include independent third-party certification, assessment of products, services or records, or even the use of such specific techniques as statistical process control (SPC). (See Chapter 8.)

Records of acceptable suppliers and subcontractors should be maintained, together with their monitored performance.

7   Customer-supplied product or services

Where a customer supplies material or services on which further transformation work is required, it is necessary to have systems that ensure the material's suitability for use and that enable monitoring and traceability of the material or service through all processes and storage. Any material that is damaged, lost, or not suitable for use should be recorded and reported to the customer. Special considerations may be necessary when the customer supplies material that is to be used in a continuous process with other purchased material. Clearly a supplier cannot be held responsible for the quality of the customer's material handled, but he is responsible for maintaining its condition while it is in his care.

8   Identification and traceability

Identification and traceability from purchased materials to finished products and services are essential if effective methods of process control are to be applied and quality problems are to be related to cause. Materials in process or bulk storage should be identified, if necessary by virtue of their location and time, and the design of procedures and record-keeping should allow for this. Traceability requirements are an optional part of an agreed specification or contract, and may be the subject of special contract conditions. In a garage, for example, a system should be developed to identify any parts removed from vehicles as either for re-use, requiring repair, etc.

9   Process control

To control the operation of any process clearly requires some planning activity, ie careful consideration of the inputs to the process so that they become suitable for the purpose. This requirement covers the core of any operation. It may be difficult to imagine, but the authors have seen too often the operation of processes about which too little is known in terms of the capability to meet the requirements.

To operate processes under controlled conditions, documented work instructions must be available to staff. These do not need to repeat the basic skills of the operator's profession, but they must contain sufficient detail to enable the process to be carried out under the specified conditions. A fully documented 'process manual' should contain, where appropriate:

(a) A description of the process, with appropriate technological information; this may be in the form of a process flowchart.
(b) A description of the plant or equipment required.
(c) Any special process 'set-up' or 'start-up' procedures.
(d) Reference to any instrumentation, calibration procedures, and measures related to control of the process.
(e) Simplified operator instruction or a summary that includes the quantity of materials required and the order in which the process is to be carried out. This may take the form of service handbooks or bulletins, etc.

For certain special processes, such as welding, plastic moulding, heat treatment, application of protective treatments, vehicle servicing, and cooking, where deficiencies may become apparent only after the product is in use, continuous monitoring of adherence to the documented procedures is the only effective method of process control.

10   Inspection and testing

All need to be either inspected or otherwise verified. The amount of inspection is clearly a function of the situation, and might consist simply of:

• Checking a product label or delivery note against a purchase order.
• Visual examination for damage in transit.
• Checking the evidence from a certificate of conformity or of analysis.

These checks are valid only if an adequate assessment of the supplier's quality system has been carried out.

Whatever the system, it must be operated in accordance with the written procedures. If bought-in materials have to be released into production before adequate verification or checking can take place, the system should ensure that it is possible to identify the material and recall it if problems arise. This may prevent the acceptance of certain bulk materials without the appropriate receiving inspection.

In process monitoring

This answers the 'Are we doing the job OK?' question, which calls for some form of process monitoring and control. Ideally it is the actual process parameters, such as temperature, cutting speed, feed rate, pressure, typing speed, flow rate, which should be monitored to ensure feedforward control of the process. The work instructions should also indicate the frequency of any in-process inspections or checks, and the action to be taken in the event of process parameters being found to be incorrect, or 'out-of-control'.

Checking finished product and/or service

Whatever final checking, inspecting or testing activities have been set out in the quality plan should be documented, including any delaying of despatch, or release of service, until the checks have been carried out. Records must be kept of all the checks, tests, measurements, etc, carried out at inwards goods or services receipt, during operation of the process, or at the final product or service stage, which are required to demonstrate conformance to the requirements or specifications. These may include certificates of conformity or analysis, and evidence on plant records or in a computer that process control parameters were actually monitored. There should also be a statement of what records are kept, for how long, and by whom.

11 Measuring, inspection and test equipment

All measuring and the test equipment relevant to the quality system must be controlled, calibrated and maintained. This includes equipments used for in-process parameter measurements and control, such as temperature and pressure gauges, as well as that used in laboratories or test/measurement areas. Where equipment is used only for observation, safety, or faulty diagnosis reasons, it may be excluded from the fully documented inspection calibration system.

The system for the instrumentation should:

(a) Refer to the measurements to be made, their accuracy and precision, and the equipment to be used to ensure the necessary capability.
(b) Identify the equipment and ensure its calibration against the appropriate standard(s) with suitable procedures, and its correct handling, preservation, storage etc.
(c) Maintain calibration records for all inspection, measuring and test equipment.
(d) Allow, where appropriate, tracking back to national standards.

12 Inspection and test status

There are essentially three statuses for all materials and services – incoming, intermediate or in-process, and finished:

- Awaiting inspection, check or test.
- Passed requirements of inspection, check or test.
- Failed requirements of inspection, check or test.

The test status of material is identified by any suitable means: labels, location, stamps markings, position in the process, records (including computers), etc. These should be used to ensure that only material or services conforming to the requirements are passed on to the next stage, or despatched. The test or check carried out may of course refer to a process-control parameter.

13 Nonconforming products or services

To prevent inadvertent use or delivery of materials or provision of services that do not conform to the specified requirements, there should be a documented system that clearly identifies and, if possible, 'segregates' them. The procedures should also show how the nonconforming output will be reworked, disposed of, accepted with concession, or regraded, and what corrective action will take place.

14   *Corrective and preventive action*

This is a very important part of the system in any organisation, since it provides the means to never ending improvement of process operation. Systematic planning is a basic requirement for effective corrective action programmes. The procedures for major corrective action should be in the form of general guidance and should define the duties of the managers, supervisors and key personnel. The detailed action to be taken will be dependent upon the circumstances prevailing at the time, and it is not therefore appropriate for the written procedures to be too detailed. All employees must be made fully aware of the general corrective action procedures appertaining to their own processes and activities. The written procedures for corrective action should be implemented when there are:

- Failures in *any* part of the quality system.
- Complaints from customers (internal or external).
- Complaints to suppliers (internal or external) and to subcontractors.

The underlying purpose of this part of the system is to eliminate the causes of nonconformance by initiation of investigations, analyses, and preventive actions. Controls must be built in to make sure that the corrective actions are taken, that they are effective, and that any necessary changes in procedures are recorded and implemented. The provision of *corrective action teams*, with regular training and updating, enables people to become used to working together to solve problems.

15   *Protection of product or service quality*

The sight of a warehouseman, in dirty boots, climbing over clean sacks of finished product to count them, still remains in one of the author's memory. A great deal of damage can be done to products and service between their 'production' and their transfer to the customer. This highlights the need for the quality system to cover such things as handling, storage, packaging, transport, and delivery of final product or services. The written procedures should be aimed at preventing damage or deterioration. The use of the correct type of packaging and labelling may invoke national or international regulations and/or codes of practice. Where contract hauliers or outside transport are employed, their ability to meet the requirements of cleanliness, schedules, etc, should be established, and appropriate procedures documented.

16   *Quality records*

The records provide objective evidence that work is being carried out in accordance with the documented procedures. Attention should be paid to identifying which records need to be retained, and to their easy retrieval. One of the author's colleagues was, on one occasion, performing a vertical audit for traceability purposes in a garment manufacturer. He selected some items from stock and, on attempting to trace back to purchased materials, discovered that final inspection records on certain items were missing. It transpired that one of the final inspectors spent 3 months in hospital, and the chief inspector, who insisted that she had stood in and carried out the necessary inspection, was too busy to write down the results. This represents a failure of the system to *demonstrate* compliance with its own requirements, and can be as serious as an ineffective procedure.

Records will have been established if the documented quality system has been set up as described above, but there must be procedures for the collection, indexing, filing or storage, retrieval, and disposition of records. Serious thought should be given to the retention time of records, which should then be stated in the documented system.

The quality records should include training and management audit and review records.

17   Quality system audits and reviews

An internal audit sets out to establish whether the quality system is being operated according to the written procedures. A *review* addresses the much wider issue of whether the quality system actually meets the requirements, and aims to determine the system's effectiveness. Clearly the results of quality audits will be used in the reviews for, if procedures are not being operated according to plan, it may be that improvements in the system are required, rather than enforcing adherence to unsuitable methods. Organisations should plan to self-police the quality system by carrying out both internal audits and reviews, and the person responsible for organising these is the manager with responsibility for co-ordinating and monitoring the whole quality system. Auditor training is now recognised as a key element in quality system implementation.

18   Training (see Chapter 12)

For *all* staff, written procedures should be established and maintained for:

- Identifying and reviewing individual training needs.
- Carrying out the training.
- Keeping records of training, including qualifications.

On-the-job training may frequently be appropriate in meeting the training requirements. It should be possible to go to the training records and establish from them objectively whether an individual has been instructed to carry out the various tasks associated with his/her job.

Training procedures may also include methods of ensuring the quality policy is understood, implemented and maintained at all levels in the organisation, for existing and new employees.

19   Servicing

If servicing is an important part of the customer requirements, eg in the provision of a burglar alarm service, procedures should be documented for its operation and to verify that it satisfies the needs. The servicing system may well include some or all of the contents of the quality system: design, documentation control, process control, training, review, etc. In particular it must ensure that:

(a) The servicing procedures are effectively carried out.
(b) Adequate resources are made available, in terms of people, time, equipment, materials, information, etc.
(c) Good interfaces exist for dealing with the customer, in terms of regular service contracts, items returned, customer complaints or call-outs.

20   Statistical techniques (see Chapter 8)

In most organisations it is necessary to measure and establish the so-called 'capability of the process'. In many industries this requires the use of certain procedures, which are grouped under the general heading of *statistical process control* (SPC).

In addition to measuring capability, controlling and improving processes through the use of statistical techniques, it might also be necessary to identify and classify lots of batches of material by their characteristics, select samples, determine any rules for acceptance or rejection of material or for adjusting the severity of inspection, and the segregation and screening of rejected materials. The system should refer to the statistical procedures used, giving the areas from their application, but always remember that statistics is simply the collection and use of data.

In considering the detailed quality system requirements, the system *must* be tailored to the needs of the 'business'. It must be seen to be an integral part of the way the business is run, and the system must be usable by all employees in never ending improvement.

5.4   Environmental management systems

Organisations of all kinds are increasingly concerned to achieve and demonstrate sound environmental performance. Many have undertaken environmental audits and review to assess this. To be effective, these need to be conducted within a structured management system, which in turn is integrated with the management activities dealing with all aspects of desired environmental performance.

Such a system should establish procedures for setting environmental policy and objectives, and achieving compliance to them. It should be designed to place emphasis on the prevention of adverse environmental effects rather than on detection after occurrence. It should also identify and assess the environmental effects arising from the organisation's existing or proposed activities, products, or services, and from incidents, accidents, and potential emergency situations. The system must identify the relevant regulatory requirements, and priorities, and pertinent environmental objectives and targets. It needs also to facilitate planning, control, monitoring, auditing and review activities to ensure that the policy is complied with, that it remains relevant, and that it is capable of evolution to suit changing circumstances.

In 1992 a British Standard BS7750 was prepared under the direction of the Environment and Pollution Standards Policy Committee in response to the increasing concerns about environmental protection and performance. It contains a specification for environmental management systems for ensuring and demonstrating compliance with stated policies and objectives. The standard is designed to enable any organisation

A cell containing a • represents a connection between the relevant sub-clauses of the two standards.

Requirement of BS EN ISO 9001 Subclause	Requirement of BS7750 Subclause										
	4.1	4.2	4.3	4.4	4.5	4.6	4.7	4.8	4.9	4.10	4.11
	Management system	Environmental policy	Organization and personnel	Environmental effects	Objectives and targets	Management programme	Manual and documentation	Operational control	Records	Audits	Reviews
4.1 Management responsibility	•	•	•							•	
4.2 Quality system	•						•				
4.3 Contract review				•	•	•					
4.4 Design control						•	•	•			
4.5 Document control							•				
4.6 Purchasing				•				•			
4.7 Purchaser supplied product				•							
4.8 Product identification									•		
4.9 Process control								•			
4.10 Inspection and testing								•			
4.11 Inspection, measuring and test equipment								•			
4.12 Inspection and test status								•			
4.13 Control of nonconforming product								•			
4.14 Corrective action								•			
4.15 Handling, storage, packaging and delivery				•				•			
4.16 Quality records									•		
4.17 Internal quality audits										•	
4.18 Training			•								
4.19 Servicing				•				•			
4.20 Statistical techniques								•			

Source – BS7750: 1992 (British Standards)

Table 5.2   *The links between BS7750: 1992 and BS ISO 9001: 1994*

to establish an effective management system as a foundation for both sound environmental performance and participation in environmental auditing schemes.

BS7750 shares common management system principles with BS ISO 9001, and organisations may elect to use an existing management system, developed in conformity with the BS ISO 9001 series, as a basis for environmental management. The new standard defines environmental policy, objectives, targets, effect, management, systems, manuals, evaluation, audits and reviews. It mirrors the ISO 9000 series requirements in many of its own eleven requirements, and it includes a guide to these in the informative Annex A.

The link to BS ISO 9001 is spelled out in a table that is reproduced in Table 5.2.

5.5 The rings of confidence

Quality systems are needed in all areas of activity, whether large or small businesses, manufacturing, service or public sector. The advantages of systems in manufacturing are obvious, but they are just as applicable in areas such as marketing, sales, personnel, finance, research and development, as well as in the service industries and public sectors.

No matter where it is implemented, a good quality system will improve process control, reduce wastage, lower costs, increase market share (or funding), facilitate training, increase staff participation, and raise morale.

The activities to be addressed in the design and implementation of a good quality-management system may be considered to be attached to a 'ring of confidence', which starts and ends with the customer (Figure 5.4). It is possible to group these into two spheres of activities:
- Those directly interacting with the customer.
- Those concerning primarily the internal activities of the supplier.

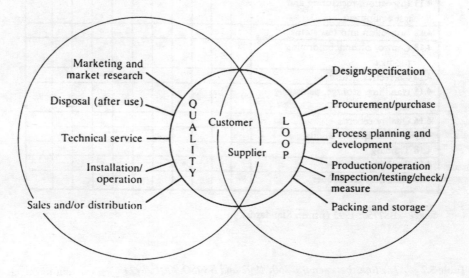

Figure 5.4   *The rings of confidence*

The overlap necessary between customer and supplier is clearly illustrated by this model. Equally obvious is that separation will lead to dysfunction and disaster.

It cannot be stated too often that the customer-supplier interactions, which generate satisfaction of needs, are just as necessary internally. The principles of quality-system design, documentation and implementation set out in this and the next chapter must apply to every single person, every department, every process transaction, and every type of organisation. The vocabulary in the engineering factory system may be different from that used in the hotel, the hospital system will be set out differently to that of the drug manufacturer, but the underlying concepts will be the same.

It is not acceptable for the managers in industries, or parts of organisations, less often associated with standards on quality systems to find 'technological' reasons for avoiding the requirements to manage quality. The authors and their colleagues have heard the excuse that 'our industry (or organisation) differs from any other industry (or organisation)', in almost every industry or organisation with which they have come into contact. Clearly there are technological differences between all industries and nearly all organisations, but in terms of managing total quality there are hardly any at all.

Senior managers in every type and size of organisation must take the responsibility for the adoption of the appropriate documented quality system. If this requires translation from 'engineering language', so be it – get someone from inside or outside the organisation to do it. Do not wait for the message to be translated into different forms – inefficiencies, waste, high costs, crippling competition, loss of market.

Chapter highlights

Why a documented system?

- An appropriate documented quality system will enable the objectives set out in the quality policy to be accomplished.
- The International Organisation for Standardisation (ISO) 9000 series set out methods by which a system can be implemented to ensure that the specified requirements are met.
- A quality system may be defined as an assembly of components, such as the organisational structure responsibilities, procedures, process, and resources.

Quality system design

- Quality systems should apply to and interact with all activities of the organisation. The activities are generally processing, communicating, and controlling. These should be documented in the form of a quality manual.
- The system should follow the PLAN DO CHECK ACT cycle, through documentation, implementation, audit and review.

Quality system requirements

- The general categories of ISO-based standards include management responsibility, quality system, contract review, design control, document and data control, purchasing, customer-supplied products or services, identification and traceability, process control, checking/measuring/inspecting of incoming materials or services, measuring/inspection/test equipment, inspection/test status, nonconforming products or services, corrective action, protection of product or service quality, quality records, quality-system audits and reviews, training, servicing and statistical techniques.

Environmental management systems

* The British Standard BS7750 contains a specification for environmental management systems for ensuring and demonstrating compliance with the stated policies and objectives, and acting as a base for auditing and review schemes. It shares common management-system principles with the ISO 9000 series.

The rings of confidence

* The activities needed in the design and implementation of a good quality system start and end with the customer, in two spheres – a customer sphere and a supplier sphere.
* Senior management in all types of industry must take responsibility for the adoption and documentation of the appropriate quality system in their organisation.

6

Quality system audit/review and self-assessment

6.1 Securing prevention by audit and review of the system

Error or defect prevention is the process of removing or controlling error/defect causes in the system. There are two major elements of this:

- Checking the system.
- Error/defect investigation and follow-up.

These have the same objectives – to find, record and report *possible* causes of error, and to recommend future corrective action.

Checking the system

There are six methods in general use:

(a) *Quality audits and reviews*, which subject each area of an organisation's activity to a systematic critical examination. Every component of the total system is included, ie, quality policy, attitudes, training, process, decision features, operating procedures, documentation. Audits and reviews, as in the field of accountancy, aim to disclose the strengths and weaknesses and the main areas of vulnerability or risk.

(b) *Quality survey*, a detailed, in-depth examination of a narrower field of activity, ie major key areas revealed by quality audits, individual plants, procedures or specific problems common to an organisation as a whole.

(c) *Quality inspection*, which takes the form of a routine scheduled inspection of a unit or department. The inspection should check standards, employee involvement and working practices, and that work is carried out in accordance with the procedures, etc.

(d) *Quality tour*, which is an unscheduled examination of a work area to ensure that, for example, the standards of operation are acceptable, obvious causes of errors are removed, and in general quality standards are maintained.

(c) *Quality sampling*, which measures by random sampling, similar to activity sampling, the error potential. Trained observers perform short tours of specific locations by prescribed routes and record the number of potential errors or defects seen. The results may be used to portray trends in the general quality situation.

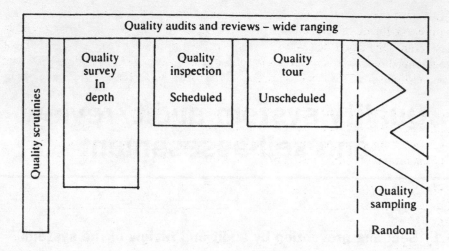

Figure 6.1   *A prevention programme combining various elements of 'checking' the system*

(f) *Quality scrutinies*, which are the application of a formal, critical examination of the process and technological intentions for new or existing facilities, or to assess the potential for mal-operation or malfunction of equipment and the consequential effects of quality. There are similarities between quality scrutinies and FMECA studies (see Chapter 8).

The design of a prevention programme, combining all these elements, is represented in Figure 6.1.

6.2 Error or defect investigations and follow-up

The investigation of errors and defects can provide valuable error prevention information. The method is based on:

- *Collecting* data and information relating to the error or defect.
- *Checking* the validity of the evidence.
- *Selecting* the evidence without making assumptions or jumping to conclusions.

The results of the analysis are then used to:

- *Decide* the most likely cause(s) of the errors or defects.
- *Notify* immediately the person(s) able to take corrective action.
- *Record* the findings and outcomes.
- *Report* them to everyone concerned, to prevent a recurrence.

The investigation should not become an inquisition to apportion blame, but focus on the positive preventive aspects. The types of follow-up to errors and their effects is shown in Table 6.1.

It is hoped that errors or defects are not normally investigated so frequently that the required skills are developed by experience, nor are these skills easily learned in a classroom. One suggested way to overcome this problem is the development of a programmed sequence of questions to form the skeleton of an error or defect investigation questionnaire. This can be set out with the following structure:

(a) *Plant equipment* – description, condition, controls, maintenance, suitability, etc.
(b) *Environment* – climatic, space, humidity, noise, etc.
(c) *People* – duties, information, supervision, instruction, training, attitudes, etc.
(d) *Systems* – procedures, instructions, monitoring, control methods, etc.

Table 6.1   *Following up errors*

System type	Aim	General effects
Investigation	To prevent a similar error or defect	*Positive:* identification notification correction
Inquisition	To identify responsibility	*Negative:* blame claims defence

6.3   Internal and external quality-system audits and reviews

A good quality system will not function without adequate audits and reviews. The system reviews, which need to be carried out periodically and systematically, are conducted to ensure that the system achieves the required effect, while audits are carried out to make sure that actual methods are adhering to the documented procedures. The reviews should use the findings of the audits, for failure to operate according to the plan often signifies difficulties in doing so. A re-examination of the procedures actually being used may lead to system improvements unobtainable by other means.

A schedule for carrying out the *audits* should be drawn up, different activities perhaps requiring different frequencies. All procedures and systems should be audited at least once during a specified cycle, but not necessarily all at the same audit. For example, every 3 months a selected random sample of work instructions and test methods could be audited, with the selection designed so that each procedure is audited at least once per year. There must be, however, a facility to adjust this on the basis of the audit results.

A quality-system *review* should be instituted, perhaps every 12 months, with the aims of:

- Ensuring that the system is achieving the desired results.
- Revealing defects or irregularities in the system.
- Indicating any necessary improvements and/or corrective actions to eliminate waste or loss.
- Checking on all levels of management.
- Uncovering potential danger areas.
- Verifying that improvements or corrective action procedures are effective.

Clearly the procedures for carrying out the audits and reviews and the results from them should be documented, and themselves be subject to review.

The assessment of a quality system against a particular standard or set of requirements by internal audit and review is known as a *first-party* assessment or approval scheme. If an *external* customer makes the assessment of a supplier against either its own or a national or international standard, a *second-party* scheme is in operation. The external assessment by an independent organisation not connected with any contract between customer and supplier, but acceptable to them both, is known as an *independent third-party* assessment scheme. The latter usually results in some form of certification or registration by the assessment body.

One advantage of the third-party schemes is that they obviate the need for customers to make their own detailed checks, saving both suppliers and customers time and money, and avoiding issues of commercial confidentiality. Just one knowledgable organisation has to be satisfied, rather than a multitude with varying levels of competence. This method often certifies suppliers for quality assurance based contracts without further checking.

Each certification body usually has its own recognised mark, which may be used by registered organisations of assessed capability in their literature, letter headings, and marketing activities. There are also publications containing lists of organisations whose quality systems and/or products and services have been assessed. To be of value, the certification body must itself be recognised and, usually, assessed and registered with a national or international accreditation scheme.

Many organisations have found that the effort of designing and implementing a written quality system good enough to stand up to external independent third-party assessment has been extremely rewarding in:

- Encouraging staff and improving morale.
- Better process control.
- Reduced wastage.
- Reduced customer service costs.

This is also true of those organisations that have obtained third-party registrations and supply companies that still insist on their own second-party assessment. The reason for this is that most of the standards on quality systems, whether national, international, or company-specific, are now very similar indeed. A system that meets the requirements of the ISO 9000 series will meet the requirements of all other standards, with only the slight modifications and small emphases here and there required for specific customers. It is the authors' experience that an assessment carried out by one of the independent certified assessment bodies is at least as rigorous and delving as any carried out by a second-part representative.

Internal system audits and reviews must be positive, and conducted as part of the preventive strategy and not as a matter of expediency resulting from quality problems. They should not be carried out only before external audits, nor should they be left to the external auditor – whether second or third party. An external auditor discovering discrepancies between actual and documented systems will be inclined to ask why the internal review methods did not discover and correct them. As this type of behaviour in financial control and auditing is commonplace, why should things be different in the control of quality?

Managements anxious to display that they are serious about quality must become fully committed to operating an effective quality system for all personnel within the organisation, not just the staff in the quality department. The system must be planned to be effective and achieve its objectives in an uncomplicated way. Having established and documented the procedures, an organisation must ensure that they are working and that everyone is operating in accordance with them. The system once established is not static; it should be flexible, to enable the constant seeking of improvements or streamlining.

Quality auditing standard

There is an International Standard Guide to quality-systems auditing (ISO 10011: 1991). This points out that audits are required to verify whether the individual elements making up quality systems are effective in achieving the stated objectives. The growing use of standards internationally emphasises the importance of auditing as a management tool for this purpose. The guidance provided in the standard can be applied equally to any one of the three specific and yet different auditing activities:

(a) *First-party or internal audits*, carried out by an organisation on its own systems, either by staff who are independent of the systems being audited, or by an outside agency.
(b) *Second-party audits*, carried out by one organisation (a purchaser or its outside agent) on another with which it either has contracts to purchase goods or services or intends to do so.
(c) *Third-party audits*, carried out by independent agencies, to provide assurance to existing and prospective customers for the product or service.

ISO 10011 covers audit objectives and responsibilities, including the roles of auditors and their independence, and those of the 'client' or auditee. It provides the following detailed guidance on audit:

- *Initiation*, including its scope and frequency.
- *Preparation*, including review of documentation, the programme, and working documents.
- *Execution*, including the opening meeting, examination and evaluation, collecting evidence, observations, and closing the meeting with the auditee.

- *Report*, including its preparation, content and distribution.
- *Completion*, including report submission and retention.

Attention is given at the end of the standard to corrective action and follow-up, where it is stressed that the improvement process should be continued by the auditee after the publication of the audit report. This may include a call by the client for a verification audit of the implementation of any corrective actions specified.

6.4 Towards a TQM standard for self-assessment

'Total quality' is the goal of many organisations, but it is difficult to find a universally accepted definition of what this actually means. For some people TQM means SPC or quality systems, for others teamwork and involvement of the workforce. Clearly there are many different views on what constitutes the 'total quality organisation' and, even with an understanding of the framework of TQM there is the difficulty of calibrating the performance or progress of any organisation towards it.

The philosophy of TQM recognises that customer satisfaction, business objectives, safety, and environment considerations are mutually dependent, and applicable in any organisation. Clearly, the application of TQM calls for investment primarily in people and time; time to implement new concepts, time to train, time for people to recognise the benefits and move forward into new or different organisational cultures. But how will organisations know when they are getting close to TQM, or whether they are even on the right road? How will they *measure* their progress?

There have been many recent developments and there will continue to be many more, in the search for a TQM standard or framework against which organisations may be assessed or measure themselves, and carry out the so-called 'gap analysis'. To many companies the ability to judge their TQM progress against an accepted set of criteria would be most valuable and informative.

Quality award criteria

Most TQM approaches strongly emphasise measurement, especially in the quality assurance and control areas. Some insist on the use of cost of quality. The recognition that total quality management is a broad culture change vehicle with internal and external focus embracing behavioural and service issues, as well as quality assurance and process control, prompted the United States to develop a widely used framework for TQM – the Malcolm Baldrige National Quality Award (MBNQA). The award itself, which is composed of two solid crystal forms over 16cm high, is presented annually to recognise companies in the USA that have 'excelled in quality management and quality achievement'. Up to two awards may be given in each of three categories: manufacturing, service, and small businesses. But it is not the award itself, or even the fact that it is presented each year by the President of the USA, which has attracted the attention of most organisations. It is the excellent framework, which is one of the closest things we have to an international standard for TQM.

The value of a structured discipline using a points scoring system has been well established in quality and safety *assurance* systems (for example, ISO 9000 Vendor Auditing). The extension of this approach to a *total* quality-auditing process has been long established in the Japanese-based 'Deming Prize', which is perhaps the most demanding and intrusive auditing process. There are other excellent models and standards used throughout the world: the British Standard BS 7850 Guide to TQM, the Marketing Quality Assurance (MQA) Specification, the UK Quality Award, and the European Quality Award.

In 1987 the MBNQA was introduced for US-based organisations. Many US companies have realised the necessity to assess themselves against the Baldrige criteria, if not to enter for the Baldrige Award then certainly as an excellent basis for self-audit and review, to highlight areas for priority attention and provide internal and external benchmarking.

The MBNQA aims to promote:

• Awareness of quality as an increasingly important element in competitiveness.
• Understanding of the requirements for quality excellence.
• Sharing of information on successful quality strategies and the benefits to be derived from their implementation.

The award criteria are built upon a set of core values and concepts:

1 Customer-driven quality.
2 Leadership.
3 Continuous improvement and learning.
4 Employee participation and development.
5 Fast response.
6 Design quality and prevention.
7 Long-range view of the future.
8 Management by fact.
9 Partnership development.
10 Corporate responsibility and citizenship.
11 Results orientation.

These are embodied in a criteria framework of seven first level categories, which are used to assess organisation. These are given in Table 6.2, along with the ten first level categories of the Deming Prize.

Figure 6.2 shows how the framework connects and integrates the categories. This has four basic elements: driver, system, measures of progress, and goal. The driver is the senior executive leadership that creates the values, goals, and systems, and guides the sustained pursuit of quality and performance objectives. The system includes a set of well-defined and designed processes for meeting the organisation's quality and performance requirements. Measures of progress provide a results-oriented basis for channelling actions to deliver ever-improving customer values and organisational performance. The goal is the basic aim of the quality process in delivering the above to the customers.

Table 6.2   *The first-level categories of the Baldrige Award and the Deming Prize*

Baldrige	Deming
1 Leadership	1 Policy
2 Information and analysis	2 Organization and management
3 Strategic planning	3 Education and dissemination
4 Human resource development and management	4 Collection, dissemination, and use of information on quality
5 Process management	5 Analysis
6 Business results	6 Standardization
7 Customer focus and satisfaction	7 Control
	8 Quality assurance
	9 Results
	10 Planning for the future

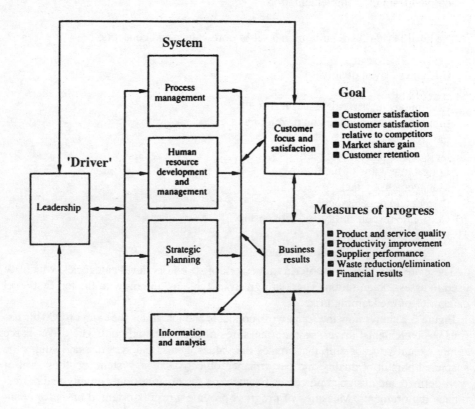

Figure 6.2   *Baldrige Award criteria framework – dynamic relationships. Source: Malcolm Baldrige National Quality Award criteria, US National Institute of Standards and Technology*

The seven criteria categories are further subdivided into examination items and areas to address. These are described in some detail in the 'Award Criteria', available from the US National Institute of Standards and Technology.

In Europe it has also been recognised that the technique of self-assessment is very useful for any organisation wishing to develop and monitor its quality culture. The European Foundation for Quality Management (EFQM) has launched a European Quality Award, which can be used effectively for a systematic review and measurement of operations. The EQA self-assessment model recognises that *processes* are the means by which a company or organisation harnesses and releases the talents of its *people* to produce *results*. Moreover, the processes and the people are the enablers which produce results.

Figure 6.3 displays graphically the principle of the European Quality Award. Essentially this states that customer satisfaction, employee satisfaction, and impact on society are achieved through leadership driving policy and strategy, people management, resources, and process, which lead ultimately to excellence in business results.

Using the European Quality Award's nine categories, it is possible to build a model of criteria and a review framework against which an organisation may face and measure itself, to examine any 'gaps'.

Many managers feel the need for a rational basis on which to measure TQM in their organisation, especially in those companies 3 or more years 'into TQM' that would like the answer to questions such as 'Where are we now?' 'Where do we need/want to be?' and 'What have we got to do to get there?' These questions need to be answered from internal employees' views, the customers' views, and the views of suppliers. A business excellence review process, which uses the EQA model criteria and lists questions that should be asked under each heading, will help any organisation to identify opportunities for improvement.

Clearly, it is necessary for any organisation to rationalise all the criteria used by the various awards. There is great overlap between them, and the main components must be the organisation's processes, quality management system, human-resource management, results and customer satisfaction.

Self-assessments provide an organisation with vital information in monitoring its progress towards its goals and total quality. The external assessments used in the

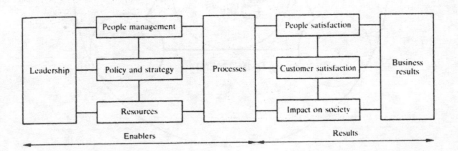
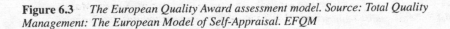

Figure 6.3   *The European Quality Award assessment model. Source: Total Quality Management: The European Model of Self-Appraisal. EFQM*

processes of making awards must be based on those self-assessments that are performed as prerequisites for improvement.

The systematic measurement and review of operations are two of the most important management activities of any TQM system. Self-assessment allows an organisation clearly to discern its strengths and areas for improvement by focusing on the relationship between the people, processes, and performance. Within any quality-conscious organisation it should be a regular activity.

6.5   Adding the systems to the TQM model

In Chapters 1 and 2 the foundations for TQM were set down. The core of total quality was established as the customer/supplier chains that extend through and out from an organisation. It was recognised that if the chains are 'cut' anywhere, processes that must be managed will be found. Within the TQM framework were identified the 'soft' outcomes of total quality, namely culture change, communication improvements and commitment.

To this foundation must be added the first hard management necessity – a quality system, based on any good international standard. This is shown in Figure 6.4.

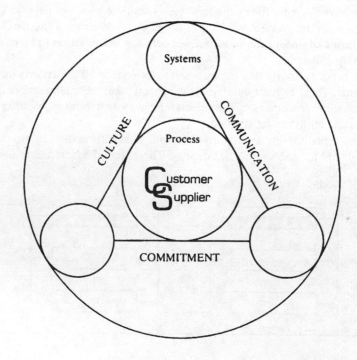

Figure 6.4   *Total quality management model – the quality system*

Chapter highlights

Securing prevention by audit and review of the system

* There are two major elements of error or defect prevention: checking the system, and error/defect investigations and follow-up. Six methods of checking quality systems are in general use: audits and review, surveys, inspections, tours, sampling and scrutinies.

Error or defect investigations and follow-up

* Investigations proceed by collecting, checking and selecting data, and analysing it by deciding causes, notifying people, recording and reporting findings and outcomes.

Internal and external quality-system audits and reviews

* A good quality system will not function without adequate audits and reviews. Audits make sure the actual methods are adhering to documented procedures. Reviews ensure the system achieves the desired effect.
* System assessment by internal audit and review is known as first-party, by external customer as second-party, and by an independent organisation as third-party certification. For the latter to be of real value the certification body must itself be recognised.

Towards a TQM standard for self-assessment

* One of the most widely used frameworks for TQM self-assessment in the USA is the Malcolm Baldrige National Quality Award (MBNQA).
* The MBNQA criteria are built on ten core values and concepts, which are embodied on a framework of seven first level categories: leadership (driver), information and analyses, strategic process planning, human resource development and management, management business (system), business results (measures of progress), and customer focus and satisfaction (goal). These are comparable with the ten categories of the Japanese Deming Prize, and the nine components of the European Quality Award: leadership, people management, policy and strategy, resources, and processes (ENABLERS), people satisfaction, customer satisfaction, impact on society, and business results (RESULTS).
* The various award criteria provide rational bases against which to measure progress towards TQM in organisations. Self-assessment against, for example, the EQA model should be a regular activity, as it identifies opportunities for improvement in performance through processes and people.

Adding the systems to the TQM model

* To the foundation framework of the customer-supplier chain, processes and the 'soft' outcomes of TQM, must be added the first hard management necessity – a quality system based on a good international standard.

Discussion questions

1 Discuss the preparations required for the negotiation of a one-year contract with a major raw material supplier.

2 Imagine that you are the chief executive or equivalent in an organisation of your own choice, and that you plan to introduce the concept of Just-in-Time (JIT) into the organisation.
(a) Prepare a briefing of your senior managers, which should include your assessment of the aims, objectives and benefits to be gained from the implementation.
(b) Outline the steps you would take to implement JIT, and explain how you would attempt to ensure its success.

3 You are the manager of a busy insurance office. Last year's abnormal winter gales led to an exceptionally high level of insurance claims for house damage caused by strong winds, and you had considerable problems in coping with the greatly increased workload. The result was excessively long delays in both acknowledging and settling customers claims.
Your area manager has asked you to outline a plan for dealing with such a situation should it arise again. The plan should identify what actions you would take to deal with the work and what, if anything, should be done now to enable you to take those actions should the need arise.
What proposals would you make, and why?

4 Explain the basic philosophy behind quality systems such as the ISO 9000 (BS 5750) series. How can an effective quality system contribute to continuous improvement in an international banking operation?

5 What role does a quality management system, based on ISO 9000 series, have in TQM? How should an organisation such as Rolls-Royce view such a standard?

6 Explain what is meant by independent third part certification to a standard such as ISO 9000, and discuss the merits of such a scheme for an organisation.

7 Compare and contrast the role of quality systems in the following organisations:
   (a) a private hospital
   (b) a medium-sized engineering company
   (c) a branch of a clearing bank

8 List the nine main categories of the European Quality Award criteria. How many such criteria should be used as the basis for a TQM self-assessment process?

9 Self-assessment using the European Quality Award (EQA) criteria enables an organisation to systematically review its business processes and results. Briefly describe the (EQA) criteria and discuss the main self-assessment methods.

10 Self-appraisal or assessment against the European Quality Award (EQA) can be used by organisations to monitor the progress of the TQM programmes.
   (a) Briefly describe the EQA criteria and explain the steps that an organisation would have to follow to carry out a self-assessment.
   (b) How could self-assessment against the EQA be used in a large multi-site organisation to drive continuous improvement?

Part Three

TQM – The Tools and the Improvement Cycle

How doth the little busy bee
Improve each shining hour
And gather honey all the day
From every opening Flower?

Isaac Watts, 1674–1748, from 'Against idleness and mischief'

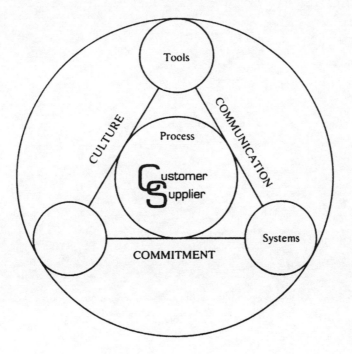

7

Measurement of Quality

7.1 Measurement and the improvement cycle

Traditionally, performance measures and indicators have been derived from cost-accounting information, often based on outdated and arbitrary principles. These provide little motivation to support attempts to introduce TQM and, in some cases, actually inhibit continuous improvement because they are unable to map process performance. In the organisation that is to survive over a long term, performance must begin to be measured by the improvements seen by the customer.

In the cycle of never ending improvement, measurement plays an important role in:

- Identifying opportunities for improvement (quality costing).
- Comparing performance against internal standards (process control and improvement).
- Comparing performance against external standards (benchmarking).

Measures are used in *process control*, eg control charts (see Chapter 8), and in *performance improvement*, eg quality improvement teams (see Chapters 10 and 11), so they should give information about how well processes and people are doing and motivate them to perform better in the future.

The authors have seen many examples of so-called performance measurement systems that frustrated improvements efforts. Various problems include systems that:

1 Produce irrelevant or misleading information.
2 Track performance in single, isolated dimensions.
3 Generate financial measures too late, eg quarterly, for mid-course corrections or remedial action.
4 Do not take account of the customer perspective, both internal and external.
5 Distort management's understanding of how effective the organisation has been in implementing its strategy.
6 Provide behaviour that *undermines* the achievement of the strategic objectives.

Typical harmful summary measures of local performance are purchase price, machine or plant efficiencies, direct labour costs, and ratios of direct to indirect labour. These are incompatible with quality-improvement measures such as process and throughput times, delivery performance, inventory reductions, and increases in flexibility, which are first and foremost *non-financial*. Financial summaries provide valuable information of course, but they should not be used for control. Effective decision-making requires direct physical measures for operational feedback and improvement.

One example of a 'measure' with these shortcomings is return on investment (ROI). ROI can be computed only after profits have been totalled for a given period. It was designed therefore as a single-period, long-term measure, but it is increasingly being used as a short-term one. Perhaps this is because most executive bonus 'packages' in the West are based on short-term measures. ROI tells us what happened, not what is happening or what will happen, and, for complex and detailed projects, ROI is inaccurate and irrelevant.

Many managers have a poor or incomplete understanding of their processes and products or services, and, looking for an alternative stimulus, become interested in financial indicators. The use of ROI, for example, for evaluating strategic requirements and performance can lead to a discriminatory allocation of resources. In many ways the financial indicators used in many businesses have remained static while the environment in which they operate has changed dramatically.

Traditionally, the measures used have not been linked to the processes where the value-adding activities take place. What has been missing is improvement measures that provide feedback to people in all areas of business operations. Of course TQM stresses the need to start with the process for fulfilling customer needs.

The critical elements of a good performance measurement and management effort look like any other list associated with total quality management:

- Leadership and commitment.
- Full employee involvement.
- Good planning.
- Sound implementation strategy.
- Measurement and evaluation.
- Control and improvement.
- Achieving and maintaining standards of excellence.

The Deming cycle of continuous improvement – PLAN DO CHECK ACT – clearly requires measurement to drive it, and yet it is a useful design aid for the measurement system itself:

PLAN: establish performance objectives and standards.
DO: measure actual performance.
CHECK: compare actual performance with the objectives and standards – determine the gap.
ACT: take the necessary actions to close the gap and make the necessary improvements.

Before we use performance measurement in the improvement cycle, however, we should attempt to answer four basic questions:

1   Why measure?
2   What to measure?
3   Where to measure?
4   How to measure?

Why measure?

It has been said often that it is not possible to manage what cannot be measured. Whether this is strictly true or not, there are clear arguments for measuring. In a quality-driven, never ending improvement environment, the following are some of the main reasons *why measurement is needed* and why it plays a key role in quality and productivity improvement.

*   To ensure customer requirements *have* been met.
*   To be able to set sensible *objectives* and comply with them.
*   To provide *standards* for establishing comparisons.
*   To provide *visibility* and provide a 'score-board' for people to *monitor* their own performance levels.
*   To highlight *quality problems* and determine which areas require *priority attention*.
*   To give an indication of the *costs of poor quality*.
*   To justify the *use of resources*.
*   To provide *feedback* for driving the improvement effort.

It is also important to know the impact of TQM on improvements in business performance, on sustaining current performance, and perhaps on reducing any decline in performance.

What to measure?

In the business of process improvement, process understanding, definition, measurement, and management are tied inextricably together. In order to assess and evaluate performance accurately, appropriate measurement must be designed, developed and maintained by people who *own* the processes concerned. They may find it necessary to measure effectiveness, efficiency, quality, impact, and productivity. In these areas there are many types of measurement, including direct output or input figures, the cost of poor quality, economic data, comments and complaints from customers, information from customer or employee surveys, etc, generally continuous variable measures (such as time) or discrete attribute measures (such as absentees).

No one can provide a generic list of what should be measured but, once it has been decided what measures are appropriate, they may be converted into indicators. These include ratios, scales, rankings, and financial and time-based indicators. Whichever measures and indicators are used by the process owners, they must reflect the true performance of the process in customer/supplier terms, and emphasise continuous improvement. Time-related measures and indicators have great value.

Where to measure?

If true measures of the effectiveness of TQM are to be obtained, there are three components that must be examined – the human, technical and business components.

The human component is clearly of major importance and the key tests are that, wherever measures are used, they must be:

1   Understood by all the people being measured.
2   Accepted by the individuals concerned.
3   Compatible with the rewards and recognition systems.
4   Designed to offer minimal opportunity for manipulation.

Technically, the measures must be the ones that truly represent the controllable aspects of the processes, rather than simple output measures that cannot be related to process management. They must also be correct, precise and accurate.

The business component requires that the measures are objective, timely, and result-oriented, and above all they must mean something to those working in and around the process, *including the customers*.

How to measure?

Measurement, as any other management system, requires the stages of design, analysis, development, evaluation, implementation and review. The system must be designed to measure *progress*, otherwise it will not engage the improvement cycle. Progress is important in five main areas: effectiveness, efficiency, productivity, quality, and impact.

Effectiveness

Effectiveness may be defined as the percentage actual output over the expected output:

$$\text{Effectiveness} = \frac{\text{Actual output}}{\text{Expected output}} \times 100 \text{ per cent}$$

Hence effectiveness looks at the *output* side of the process and is about the implementation of the objectives – doing what you said you would do. Effectiveness measures should reflect whether the organisation, group or process owner(s) are achieving the desired results, accomplishing the right things. Measures of this may include:

*   Quality, eg a grade of product, or a level of service.
*   Quantity, eg tonnes, lots, bedrooms cleaned, accounts opened.
*   Timeliness, eg speed of response, product lead times, cycle time.
*   Cost/price, eg unit costs.

Efficiency

Efficiency is concerned with the percentage resources actually used over the resources that were planned to be used:

$$\text{Efficiency} = \frac{\text{Resources actually used}}{\text{Resources planned to be used}} \times 100 \text{ per cent}$$

Clearly, this is a process *input* issue and measures performance of the process system management. It is, of course, possible to use resources 'efficiently' while being *ineffective*, so performance efficiency improvement must be related to certain output objectives.

All process inputs may be subjected to efficiency measurement, so we may use labour/staff efficiency, equipment efficiency (or utilisation), material efficiency, information efficiency, etc. Inventory data and throughput times are often used in efficiency and productivity ratios.

Productivity

Productivity measures should be designed to relate the process outputs to its inputs:

$$\text{Productivity} = \frac{\text{Outputs}}{\text{Inputs}}$$

and this may be quoted as expected or actual productivity:

$$\text{Expected productivity} = \frac{\text{Expected output}}{\text{Resources expected to be consumed}}$$

$$\text{Actual productivity} = \frac{\text{Actual output}}{\text{Resources actually consumed}}$$

There is a vast literature on productivity and its measurement, but simple ratios such as tonnes per man-hour (expected and actual), pages of word-processing per operator-day, and many others like this are in use. Productivity measures may be developed for each input or a combination of inputs, eg sales/all employee costs.

Quality

This has been defined elsewhere of course (see Chapter 1). The *non-quality* related measures include the simple counts of defect or error rates (perhaps in parts per million), percentage outside specification or Cp/Cpk values, deliveries not on time, or more generally as the costs of poor quality. When the positive costs of prevention of poor quality are included, these provide a balanced measure of the costs of quality.

The quality measures should also indicate positively whether we are doing a good job in terms of customer satisfaction, implementing the objectives, and whether the designs, systems, and solutions to problems are meeting the requirements. These really are voice-of-the customer measures.

Impact

Impact measures should lead to key performance indicators for the business or organisation, including monitoring improvement over time. Value-added management (VAM) requires the identification and elimination of all non-value-adding wastes, including time. Value added is simply the volume of sales (or other measure of 'turnover') minus the total input costs, and provides a good direct measure of the impact of the improvement process on the performance of the business. A related ratio, percentage return on value added (ROVA):

$$ROVA = \frac{\text{Net profits before tax}}{\text{Value added}} \times 100 \text{ per cent}$$

is another financial indicator that may be used.

Other measures or indicators of impact on the business are *growth* in sales, assets, numbers of passengers/students, etc, and *asset-utilisation* measures such as return on investment (ROI) or capital employed (ROCE), earnings per share, etc.

Some of the impact measures may be converted to people productivity ratios, eg:

$$\frac{\text{Value added}}{\text{Number of employees (or employee costs)}}$$

Activity-based costing (ABC) is an information system that maintains and processes data on an organisation's activities and cost objectives. It is based on the activities performed being identified and the costs being traced to them. ABC uses various 'cost drivers' to trace the cost of activities to the cost of the products or services. The activity and cost-driver concepts are the heart of ABC. Cost drivers reflect the demands placed on activities by products, services or other cost targets. Activities are processes or procedures that cause work and thereby consume resources. This clearly measures impact, both on and by the organisation.

7.2 The implementation of performance measurement systems

It has already been established that a good measurement system will start with the customer and measure the right things. The value of any measure clearly needs to be compared with the cost of producing it. There will be appropriate measures for different parts of the organisation, but everywhere they must relate process performance to the needs of the process customer. All critical parts of the process must be measured, and it is often better to start with simple measures and improve them.

There must be a recognition of the need to distinguish between different measures for different purposes. For example, an operator may measure time, various process parameters, and amounts, while at the management level measuring costs and delivery timeliness may be more appropriate.

Participation in the development of measures enhances their understanding and acceptance. Process-owners can assist in defining the required performance measures, provided that senior managers have communicated their mission clearly, defined the critical success factors, and identified the critical processes (see Chapter 13).

If all employees participate, and own the measurement processes, there will be lower resistance to the system, and a positive commitment towards future changes will be engaged. This will derive from the 'volunteered accountability', which will in turn make the individual contribution more visible. Involvement in measurement also strengthens the links in the customer-supplier chains and gives quality improvement teams much clearer objectives. This should lead to greater short-term and long-term productivity gains.

There are a number of possible reasons why measurement systems fail:

1   They do not define performance operationally.
2   They do not relate performance to the process.
3   The boundaries of the process are not defined.
4   The measures are misunderstood or misused or measure the wrong things.
5   There is no distinction between control and improvement.
6   There is a fear of exposing poor and good performance.
7   It is seen as an extra burden in terms of time and reporting.
8   There is a perception of reduced autonomy.
9   Too many measurements are focused internally and too few are focused externally.
10 There is a fear of the introduction of tighter management controls.

These and other problems are frequently due to poor planning at the implementation stage or a failure to assess current systems of measurement. Before the introduction of a total quality-based performance measurement system, an audit of the existing systems should be carried out. Its purpose is to establish the effectiveness of existing measures, their compatibility with the quality drive, their relationship with the processes concerned, and their closeness to the objectives of meeting customer requirements. The audit should also highlight areas where performance has not been measured previously, and indicate the degree of understanding and participation of the employees in the existing systems and the actions that result.

Generic questions that may be asked during the audit include:

• Is there a performance measurement system in use?
• Has it been effectively communicated throughout the organisation?
• Is it systematic?
• Is it efficient?
• Is it well understood?
• Is it applied?
• Is it linked to the mission and objectives of the organisation?
• Is there a regular review and update?
• Is action taken to improve performance following the measurement?
• Are the people who own the processes engaged in measuring their own performance?
• Have employees been properly trained to conduct the measurement?

Following such an audit, there are twelve basic steps for the introduction of TQM-based performance measurement. Half of these are planning steps and the other half implementation.

Planning

1   Identify the purpose of conducting measurement, ie is it for:
   (a) Reporting, eg ROI reported to shareholders.
   (b) Controlling, eg using process data on control charts.
   (c) Improving, eg monitoring the results of a quality improvement team project.

2   Choose the right balance between individual measures (activity- or task-related) and group measures (process- and sub-process-related) and make sure they reflect process performance.
3   Plan to measure all the key elements of performance, not just one, eg time, cost, and product quality variables may all be important.
4   Ensure that the measures will reflect the voice of the internal/external customers.
5   Carefully select measures that will be used to establish standards of performance.
6   Allow time for the learning process during the introduction of a new measurement system

Implementation

7   Ensure full participation during the introductory period and allow the system to mould through participation.
8   Carry out cost/benefit analysis on the data generation, and ensure measures that have high 'leverage' are selected.
9   Make the effort to spread the measurement system as widely as possible, since effective decision-making will be based on measures from *all* areas of the business operation.
10 Use *surrogate* measures for subjective areas where quantification is difficult, eg improvements in morale may be 'measured' by reductions in absenteeism or staff turnover rates.
11 Design the measurement systems to be as flexible as possible, to allow for changes in strategic direction and continual review.
12 Ensure that the measures reflect the quality drive by showing small incremental achievements that match the never ending improvement approach.

In summary the measurement system must be designed, planned and implemented to reflect customer requirements, give visibility to the processes and the progress made, communicate the total quality effort and engage the never ending improvement cycle. So it must itself be periodically reviewed.

7.3 Benchmarking

Product, service and process improvements can take place only in relation to established standards, and the improvements then being incorporated into the new standards. *Benchmarking*, one of the most transferable aspects of Rank Xerox's approach to total quality management, and thought to have originated in Japan, measures an organisation's operations, products and services against those of its competitors in a ruthless fashion. It is a means by which targets, priorities and operations that will lead to competitive advantage can be established.

Benchmarking is the continuous process of measuring products, services and processes against those of industry leaders or the toughest competitors. This results in a search for best practice, those that will lead to superior performance, through measuring performance, continuously implementing change, and emulating the best.

There may be many reasons for carrying out benchmarking. Some of them are set against various objectives in Table 7.1. The links between benchmarking and TQM are clear – establishing objectives based on industry best practice should directly contribute to better meeting of the internal and external customer requirements.

There are four basic types of benchmarking:

Internal – a comparison of internal operations.
Competitive – specific competitor to competitor comparisons for a product or function of interest.
Functional – comparisons to similar functions within the same broad industry or to industry leaders.
Generic – comparisons of business processes that are very similar, regardless of the industry.

Table 7.1 *Reasons for benchmarking*

Objectives	Without benchmarking	With benchmarking
Becoming competitive	• Internally focused • Evolutionary change	• Understanding of competitiveness • Ideas from proven practices
Industry best practices	• Few solutions • Frantic catch up activity	• Many options • Superior performance
Defining customer requirements	• Based on history or gut feeling • Perception	• Market reality • Objective evaluation
Establishing effective goals and objectives	• Lacking external focus • Reactive	• Credible, unarguable • Proactive
Developing true measures of productivity	• Pursuing pet projects • Strength and weaknesses not understood • Route of least resistance	• Solving real problems • Understanding outputs • Based on industry best practices

The evolution of benchmarking in an organisation is likely to progress through four focuses. Initially attention will be concentrated on competitive products or services, including, for example, design, development and operational features. This should develop into a focus on industry best practices and may include, for example, aspects of distribution or service. The real breakthrough is when the organisation focuses on all aspects of the total business performance, across all functions and aspects, and addresses current *and projected* performance gaps. This should lead to the final focus on true continuous improvement.

At its simplest competitive benchmarking, the most common form, requires every department to examine itself against its counterpart in the best competing companies. This includes a scrutiny of all aspects of their activities. Benchmarks that may be important for *customer satisfaction*, for example, might include:

- Product or service consistency.
- Correct and on-time delivery.
- Speed of response or new product development.
- Correct billing.

For *impact* the benchmarks may be:

- Waste, rejects or errors.
- Inventory levels/work in progress.
- Costs of operation.
- Staff turnover.

The task is to work out what has to be done to improve on the competition's performance in each of the chosen areas.

At regular (say, weekly) meetings, managers should discuss the results of the competitive benchmarking, and on a daily basis departmental managers should discuss quality problems with staff. One afternoon may be set aside for the benchmark meetings, followed by a 'walkabout', when the manager observes the activities actually taking place and compares them mentally with the competitors' operations.

The process has fifteen stages, and these are all focused on trying to *measure* comparisons of competitiveness:

PLAN

Select department(s) or process group(s) for benchmarking.
Identify best competitor, perhaps using customer feed-back or industry observers.
Identify benchmarks.
Bring together the appropriate team.
Decide information and data-collection methodology (do not forget desk research!)
Prepare for any visits and interact with target organisations.
Use data-collection methodology.

ANALYSE

Compare the organisation and its 'competitors', using the benchmark data.
Catalogue the information and create a 'competency centre'.
Understand the 'enabling processes' as well as the performance measures.

DEVELOP

Set new performance level objectives/standards.
Develop action plans to achieve goals and integrate into the organisation.

IMPROVE   Implement specific actions and integrate them into the business processes.

REVIEW   Monitor the results and improvements.
                 Review the benchmarks and the ongoing relationship with the target organisation.

Benchmarking is very important in the administrative areas, since it continuously measures services and practices against the equivalent operation in the toughest direct competitors or organisations renowned as leaders in the areas, even if they are in the same organisation. An example of quantitative benchmarks in absenteeism is given in Table 7.2.

Technologies and conditions vary between different industries and markets, but the basic concepts of measurement and benchmarking are of general validity. The objective should be to produce products and services that conform to the requirements of the customer in a never ending improvement environment. The way to accomplish this is to use the continuous improvement cycle in all the operating departments – nobody should be exempt. Measurement and benchmarking are not separate sciences or unique theories of quality management, but rather strategic approaches to getting the best out of people, processes, products, plant, and programmes.

Table 7.2 *Quantitative benchmarking in absenteeism*

Organization's absence level (%)	Productivity opportunity
Under 3	This level matches an aggressive benchmark that has been achieved in 'excellent' organizations.
3-4	This level may be viewed within the organization as a good performance – representing a moderate productivity opportunity improvement.
5-8	This level is tolerated by many organizations but represents a major improvement opportunity.
9-10	This level indicates that a serious absenteeism problem exists.
Over 10	This level of absenteeism is totally unacceptable.

7.4 Costs of quality

Manufacturing a quality product, providing a quality service, or doing a quality job – one with a high degree of customer satisfaction – is not enough. The cost of achieving these goals must be carefully managed, so that the long-term effect of quality costs on the business or organisation is a desirable one. These costs are a true measure of the quality effort. A competitive product or service based on a balance between quality and cost factors is the principal goal of responsible management. The objective is best accomplished with the aid of competent analysis of the costs of quality (COQ).

The analysis of quality related costs is a significant management tool that provides:

- A method of assessing the effectiveness of the management of quality.
- A means of determining problem areas, opportunities, savings, and action priorities.

The costs of quality are no different from any other costs. Like the costs of maintenance, design, sales, production/operations, and other activities, they can be budgeted, measured and analysed.

Having specified the quality of design, the operating units have the task of matching it. The necessary activities will incur costs that may be separated into prevention costs, appraisal costs and failure costs, the so-called P–A–F model first presented by Feigenbaum. Failure costs can be further split into those resulting from internal and external failure.

Prevention costs

These are associated with the design, implementation and maintenance of the total quality management system. Prevention costs are planned and are incurred before actual operation. Prevention includes:

Product or service requirements

The determination of requirements and the setting of corresponding specifications (which also takes account of process capability) for incoming materials, processes, intermediates, finished products and services.

Quality planning

The creation of quality, reliability, and operational, production, supervision, process control, inspection and other special plans, eg pre-production trials, required to achieve the quality objective.

Quality assurance

The creation and maintenance of the quality system.

Inspection equipment

The design, development and/or purchase of equipment for use in inspection work.

Training

The development, preparation and maintenance of training programmes for operators, supervisors, staff, and managers both to achieve and maintain capability.

Miscellaneous

Clerical, travel, supply, shipping, communications and other general office management activities associated with quality.

Resources devoted to prevention give rise to the '*costs of doing it right the first time*'.

Appraisal costs

These costs are associated with the supplier's and customer's evaluation of purchased materials, processes, intermediates, products and services to assure conformance with the specified requirements. Appraisal includes:

Verification

Checking of incoming material, process set-up, first-offs, running processes, intermediates and final products, including product or service performance appraisal against agreed specifications.

Quality audits

To check that the quality system is functioning satisfactorily.

Inspection equipment

The calibration and maintenance of equipment used in all inspection activities.

Vendor rating

The assessment and approval of all suppliers, of both products and services.

Appraisal activities result in the '*costs of checking it is right*'.

Internal failure costs

These costs occur when the results of work fail to reach designed quality standards and are detected before transfer to the customer takes place. Internal failure includes the following:

Waste

The activities associated with doing unnecessary work or holding stocks as the result of errors, poor organisation or poor communications, the wrong materials, etc.

Scrap

Defective product, material or stationery that cannot be repaired, used or sold.

Rework or rectification

The correction of defective material or errors to meet the requirements.

Re-inspection

The re-examination of products or work that have been rectified.

Downgrading

A product that is usable but does not meet specifications may be downgraded and sold as 'second quality' at a low price.

Failure analysis

The activity required to establish the causes of internal product or service failure.

External failure costs

These costs occur when products or services fail to reach design quality standards but are not detected until after transfer to the consumer. External failure includes:

Repair and servicing

Either of returned products or those in the field.

Warranty claims

Failed products that are replaced or services re-performed under some form of guarantee.

Complaints

All work and costs associated with handling and servicing of customers' complaints.

Returns

The handling and investigation of rejected or recalled products or materials, including transport costs.

Liability

The result of product or service liability litigation and other claims, which may include a change of contract.

Loss of goodwill

The impact on reputation and image, which impinges directly on future prospects for sales.

External and internal failures produce the '*costs of getting it wrong*'.

Order re-entry, retyping, unnecessary travel and telephone calls, conflict, are just a few examples of the wastage or failure costs often excluded. Every organisation must be aware of the costs of getting it wrong, and management needs to obtain some idea how much failure is costing each year.

Clearly, this classification of cost elements may be used to interrogate any internal transformation process. Using the internal customer requirements concept as the standard for failure, these cost assessments can be made wherever information, data, materials, service or artefacts are transferred from one person or one department to

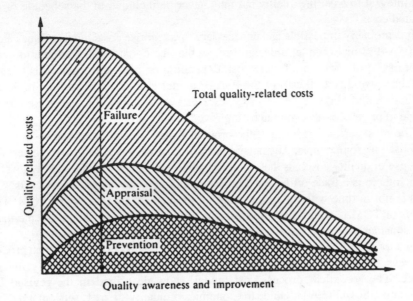

Figure 7.1 *Increasing quality awareness and improvement activities. Source: British Standard BS6143, 1991*

another. It is the 'internal' costs of lack of quality that lead to the claim that approximately one-third of *all* our efforts are wasted.

The relationship between the quality related costs of prevention, appraisal, and failure and increasing quality awareness and improvement in the organisation is shown in Figure 7.1. Where the quality awareness is low the total quality related costs are high, the failure costs predominating. As awareness of the cost to the organisation of failure gets off the ground, through initial investment in training, an increase in appraisal costs usually results. As the increased appraisal leads to investigations and further awareness, further investment in prevention is made to improve design features, processes and systems. As the preventive action takes effect, the failure *and* appraisal costs fall and the total costs reduce.

The first presentations of the P–A–F model suggested that there may be an optimum operating level at which the combined costs are at the minimum. The authors, however, have not yet found one organisation in which the total costs have risen following investment in prevention.

7.5 The process model for quality costing

The P–A–F model for quality costing has a number of drawbacks. In TQM, prevention of problems, defects, errors, waste, etc, is one of the prime functions, but it can be argued that everything a well managed organisation does is directed at preventing quality problems. This makes separation of *prevention costs* very difficult. There are clearly a range of prevention activities in any manufacturing or service organisation that are integral to ensuring quality but may never be included in the schedule of quality related costs.

It is probably impossible and unnecessary to categorise costs into the three categories of P–A–F. For example, a design review may be considered a prevention cost, an appraisal cost, or even a failure cost, depending on how and where it is used in the process. Another criticism of the P–A–F model is that it focuses attention on cost reduction and plays down, or in some cases even ignores, the positive contribution made to price and sales volume by improved quality.

The most serious criticism of the original P–A–F model presented by Feigenbaum and used in, for example, Australian Standard 2561 (1982) 'Guide to the determination and use of quality costs', is that it implies an acceptable 'optimum' quality level above which there is a trade-off between investment in prevention and failure costs. Clearly, this is not in tune with the never ending improvement philosophy of TQM. The key focus of TQM is on process improvement, and a cost categorisation scheme that does not consider process costs, such as the P–A–F model, has limitations.

In a total quality related cost system that focuses on processes rather than products or services, the operating costs of generating customer satisfaction will be of prime importance. The so-called 'process cost model', now described in the revised British Standard BS6143 (1991) 'Guide to economics of quality', Part 1, sets out a method for applying quality costing to any process or service. It recognises the importance of process ownership and measurement, and uses process modelling to simplify classification. The categories of the cost of quality (COQ) have been rationalised into the cost of conformance (COC) and the cost of non-conformance (CONC):

$$COQ = COC + CONC$$

The cost of conformance (COC) is the process cost of providing products or services to the required standards, by a given specified process in the most effective manner, ie the cost of the ideal process where every activity is carried out according to the requirements first time, every time. The cost of nonconformance (CONC) is the failure cost associated with a process not being operated to the requirements, or the cost due to variability in the process. Part 2 of BS 6143 (1991) still deals with the P–A–F model, but without the 'optimum'/minimum cost theory (see Figure 7.1).

Process cost models can be used for any process within an organisation and developed for the process by flowcharting. This will identify the key process steps and the parameters that are monitored in the process. The process cost elements should then be identified and recorded under the categories of product/service (outputs), and people, systems, plant or equipment, materials, environment, information (inputs). The COC and CONC for each stage of the process will comprise a list of all the parameters monitored.

Steps in process cost modelling

Process cost modelling is a methodology that lends itself to stepwise analysis, and the following are the key stages in building the model.

1   Choose a key process to be analysed, identify and name it, eg Retrieval of Medical Records (Acute Admissions).
2   Define the process and its boundaries.

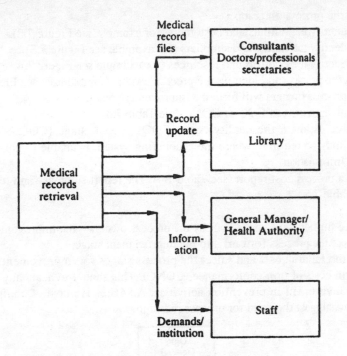

Figure 7.2   *Building the model: identify outputs and customers*

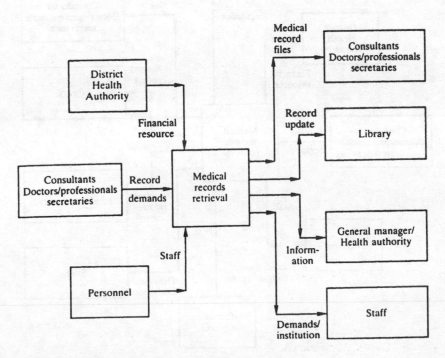

Figure 7.3   *Building the model: identify inputs and suppliers*

3   Construct the process diagram:
  (a)   Identify the outputs and customers (for example see Figure 7.2).
  (b)   Identify the inputs and suppliers (for example see Figure 7.3).
  (c)   Identify the controls and resources (for example see Figure 7.4).
4   Flowchart the process and identify the process owners (for example see Figure 7.5). Note, the process owners will form the improvement team.
5   Allocate the activities as COC or CONC (see Table 7.3).
6   Calculate or estimate the quality costs (COQ) at each stage (COC + CONC). Estimates may be required where the accounting system is unable to generate the necessary information.
7   Construct a process cost report (see Table 7.4). The report summary and results are given in Table 7.5.

There are three further steps carried out by the process owners – the improvement team – which take the process forward into the improvement stage:
8   Prioritise the failure costs and select the process stages for improvement through reduction in costs of non-conformance (CONC). This should indicate any requirements for investment in prevention activities. An excessive cost of conformance (COC) may suggest the need for process redesign.

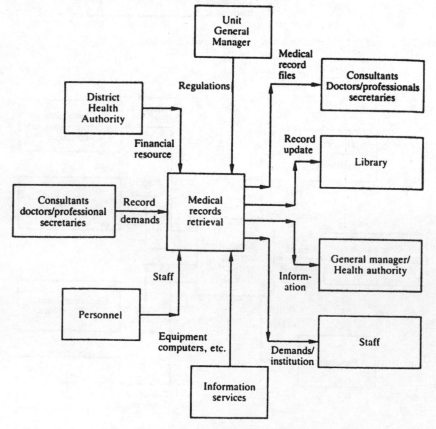

Figure 7.4   *Building the model: identify controls and resources*

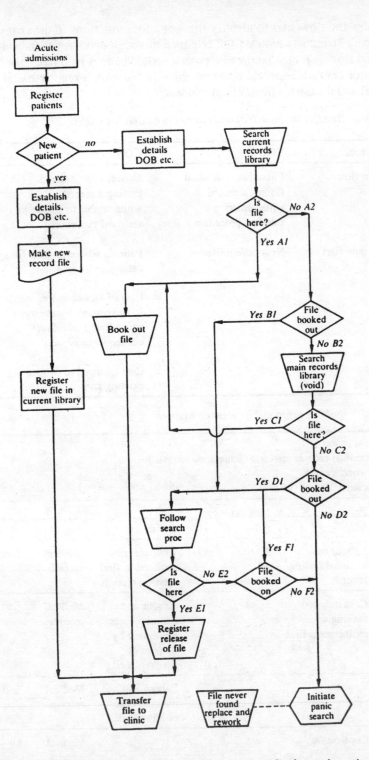

Figure 7.5   *Present practice flowchart for acute admissions, medical records retrieval*

9 Review the flowchart to identify the scope for reductions in the cost of conformance. Attempts to reduce COC require a thorough process understanding, and a second flowchart of what the new process should be may help (see Chapter 4).

10 Monitor conformance and non-conformance costs on a regular basis, using the model and review for further improvements.

Table 7.3   *Building the model: allocate activities as COC or CONC*

Key activities	COC	CONC
Search for files	Labour cost incurred finding a record while adhering to standard procedure	Labour cost incurred finding a record while unable to adhere to standard procedure
Make up new files	New patient files	Patients whose original files cannot be located
Rework		Cost of labour and materials for all rework files/records never found as a direct consequence of . . .
Duplication		Cost incurred in duplicating existing files

Table 7.4   *Building the model: process cost report*

Process cost report
Process: medical records retrieval (acute admissions)
Process owner: various
Time allocation: 4 days (96 hrs)

Process COC	Process CONC	Cost details Act	Synth	Definition	Source	
	Labour cost incurred finding records	# ref. sample		Cost of time required to find missing records	Medical records	$98
	Cost incurred making up replacement files		#	Labour and material costs multiplied by number of files replaced	Medical records	$40
	Rework		#	Labour and material cost of all rework	Medical records	$50
	Duplication		#		Medical records	$9

The process cost model approach must be seen as more than a simple tool to measure the financial implications of the gap between the actual and potential performance of a process. The emphasis given to the process, improving the understanding, and seeing in detail where the costs occur, should be an integral part of quality improvement.

Table 7.5   *Process cost model: report summary*

Labour cost
> 13.75 hrs x $5.80/hr = $80
> $80 x overhead and contribution factor 22%
> = $98

Replacement costs
> No of files unfound 9
> Cost to replace each file $4.50
> Overall cost $40

Rework costs
> 2 x Pathology reports to be retyped $50

Duplication costs
> No of files duplicated 2
> Cost per file $4.50
> Overall cost $9

TOTAL COST $197

RESULTS
Acute admissions operated 24 hrs/day 365 days/year
This project established a cost of nonconformance of approx. $197
This equates to $197 x 365/4 = $17,976
or two personnel fully employed for 12 months

Acknowledgment

The authors are grateful to their colleagues Dr Mohamed Zairi and Dr Les Porter, at the European Centre for TQM, Bradford, UK, for their significant contribution to this chapter.

Chapter highlights

Measurement and the improvement cycle

- Traditional performance measures based on cost-accounting information provide little to support TQM, because they do not map process performance and improvements seen by the customer.

- Measurement is important in identifying opportunities, and comparing performance internally and externally. Measures, typically non-financial, are used in process control and performance improvement.
- Some financial indicators, such as ROI, are often inaccurate, irrelevant and too late to be used as measures for performance improvement.
- The Deming cycle of PLAN DO CHECK ACT is a useful design aid for measurement systems, but firstly four basic questions about measurement should be asked, ie why, what, where, and how.
- In answering the question 'how to measure?' progress is important in five main areas: effectiveness, efficiency, productivity, quality, and impact.
- Activity-based costing (ABC) is based on the activities performed being identified and costs traced to them. ABC uses cost drivers, which reflect the demands placed on activities.

The implementation of performance measurement systems

- The value of any measure must be compared with the cost of producing it. All critical parts of the process must be measured, but it is often better to start with the simple measures and improve them.
- Process-owners should take part in defining the performance measures, which must reflect customer requirements.
- Prior to introducing TQM measurement, an audit of existing systems should be carried out to establish their effectiveness, compatibility, relationship and closeness to the customer.
- Following the audit, there are twelve basic steps for implementation, six of which are planning steps.

Benchmarking

- Benchmarking measures an organisation's operations, products, and services against those of its competitors. It will establish targets, priorities, and operations, leading to competitive advantage.
- There are four basic types of benchmarking: internal, competitive, functional, and generic. The evolution of benchmarking in an organisation is likely to progress through four focuses towards continuous improvement.
- The implementation of benchmarking has fifteen stages, which are categorised into plan, analyse, develop, improve, and review.

Costs of quality

- A competitive product or service based on a balance between quality and cost factors is the principal goal of responsible management.
- The analysis of quality related costs provide a method of assessing the effectiveness of the management of quality and of determining problem areas, opportunities, savings, and action priorities.
- Total quality costs may be categorised into prevention, appraisal, internal failure, and external failure costs, the P–A–F model.

- Prevention costs are associated with doing it right the first time, appraisal costs with checking it is right, and failure costs with getting it wrong.
- When quality awareness in an organisation is low, the total quality related costs are high, the failure costs predominating. After an initial rise in costs, mainly through the investment in training and appraisal, increasing investment in prevention causes failure, appraisal and total costs to fall.

The process model for quality costing

- The P–A–F model for quality costing has a number of drawbacks, mainly due to estimating the prevention costs, and its association with an 'optimised' or minimum total cost.
- An alternative – the process cost model – rationalises costs of quality (COQ) into the cost of conformance (COC) and the cost of non-conformance (CONC). COQ = COC + CONC at each process stage.
- Process cost modelling calls for choice of a process and its definition; construction of a process diagram; identification of outputs and customers, inputs and suppliers, controls and resources; flowcharting the process and identifying owners; allocating activities as COC or CONC; and calculating the costs. A process cost report with summaries and results is produced.
- The failure costs or CONC should be prioritised for improvements.

8

Tools and techniques for quality improvement

8.1 A systematic approach

In the never-ending quest for improvement in the ways processes are operated, numbers and information will form the basis for understanding, decisions and actions; and a thorough data gathering, recording and presentation system is essential.

In addition to the basic elements of a quality system that provide a framework for recording, there exists a set of methods the Japanese quality guru Ishikawa has called the seven basic tools. These should be used to interpret and derive the maximum use from data. The simple methods listed below, of which there are clearly more than seven, will offer any organisation means of collecting, presenting, and analysing most of its data:

- Process flowcharting – what is done?
- Check sheets/tally charts – how often is it done?
- Histograms – what do overall variations look like?
- Scatter diagrams – what are the relationships between factors?
- Stratification – how is the data made up?
- Pareto analysis – which are the big problems?
- Cause and effect analysis and Brainstorming (including CEDAC, NGT, and the five whys) – what causes the problems?
- Force-field analysis – what will obstruct or help the change or solution?
- Emphasis curve – which are the most important factors?
- Control charts – which variations to control and how?

Sometimes more sophisticated techniques, such as analysis of variance, regression analysis, and design of experiments, need to be employed.

The effective use of the tools requires their application by the people who actually work on the processes. Their commitment to this will be possible only if they are assured that management cares about improving quality. Managers must show they are serious by establishing a systematic approach and providing the training and implementation support required.

Improvements cannot be achieved without specific opportunities, commonly called problems, being identified or recognised. A focus on improvement opportunities leads to the creation of teams whose membership is determined by their work on and detailed knowledge of the process, and their ability to take improvement action. The teams must then be provided with good leadership and the right tools to tackle the job.

The systematic approach (Figure 8.1) should lead to the use of factual information, collected and presented by means of proven techniques, to open a channel of communications not available to the many organisations that do no follow this or a similar approach to problem solving and improvement. Continuous improvements in the quality of products, services, and processes can often be obtained without major capital investment, if an organisation marshals its resources, through an understanding and breakdown of its processes in this way.

By using reliable methods, creating a favourable environment for team based problem solving, and continuing to improve using systematic techniques, the never-ending improvement helix (see Chapter 2) will be engaged. This approach demands the real time management of data, and actions on processes and inputs, not outputs. It will require a change in the language of many organisations from percentage defects, percentage 'prime' product, and number of errors, to *process capability*. The climate must change from the traditional approach of 'If it meets the specification, there are no problems and no further improvements are necessary'. The driving force for this will be the need for better internal and external customer satisfaction levels, which will lead to the continuous improvement question, 'Could we do the job better?'

8.2   Some basic tools and techniques

Understanding processes so that they can be improved by means of the systematic approach requires knowledge of a simple kit of tools or techniques. What follows is a brief description of each technique, but a full description and further examples of some of them may be found in references 1 and 2 (p 142).

Process flowcharting

The use of this technique, which is described in Chapter 4, ensures a full understanding of the inputs and flow of the process. Without that understanding, it is not possible to draw the correct flowchart of the process. In flowcharting it is important to remember that in all but the smallest tasks no single person is able to complete a chart without help from others. This makes flowcharting a powerful team forming exercise.

Check sheets or tally charts

A check sheet is a tool for data gathering, and a logical point to start in most process control or problem solving efforts. It is particularly useful for recording direct observations and helping to gather in facts rather than opinions about the process. In the recording process it is essential to understand the difference between data and numbers.

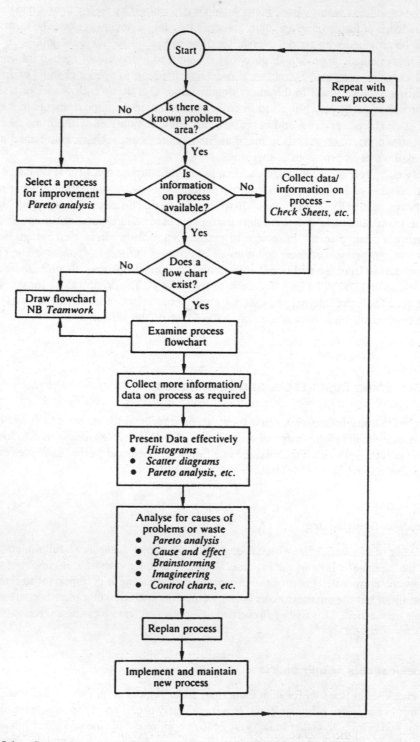

Figure 8.1 *Strategy for process improvement*

Observer F. Oldsman		Computer No. 148		Date 26 June
Number of observations 95			Total	Percentage
Computer in use		⅃⅃⅂⅂ ⅃⅃⅂⅂ ⅃⅃⅂⅂ ⅃⅃⅂⅂ ⅃⅃⅂⅂ ⅃⅃⅂⅂ ⅃⅃⅂⅂ ⅃⅃⅂⅂ ⅃⅃⅂⅂ ⅃⅃⅂⅂ ⅃⅃⅂⅂	55	57· 9
Computer idle	Repairs	⅃⅃⅂⅂	5	5·3
	No work	⅃⅃⅂⅂ ⅃⅃⅂⅂ ‖	12	12·6
	Operator absent	⅃⅃⅂⅂ ⅃⅃⅂⅂	10	10·5
	System failure	⅃⅃⅂⅂ ⅃⅃⅂⅂ ‖‖	13	13·7

Figure 8.2   *Activity sampling record in an office*

Data are pieces of information, including numerical information, that are useful in solving problems, or provide knowledge about the state of a process. Numbers alone often represent meaningless measurements or counts, which tend to confuse rather than to enlighten. Numerical data on quality will arise either from counting or measurement.

The use of simple check sheets or tally charts aids the collection of data of the right type, in the right form, at the right time. The objectives of the data collection will determine the design of the record sheet used. An example of a tally chart is shown in Figure 8.2. this gives rise to a frequency distribution.

Histograms

Histograms show, in a very clear pictorial way, the frequency with which a certain value or group of values occurs. They can be used to display both attribute and variable data, and are an effective means of letting the people who operate the process know the results of their efforts. Data gathered on truck turnround times is drawn as a histogram in Figure 8.3.

Scatter diagrams

Depending on the technology, it is frequently useful to establish the association, if any, between two parameters or factors. A technique to begin such an analysis is a simple X–Y plot of the two sets of data. The resulting grouping of points on scatter diagrams (eg Figure 8.4) will reveal whether or not a strong or weak, positive or negative, correlation exists between the parameters. The diagrams are simple to construct and easy to interpret, and the absence of correlation can be as revealing as finding that a relationship exists.

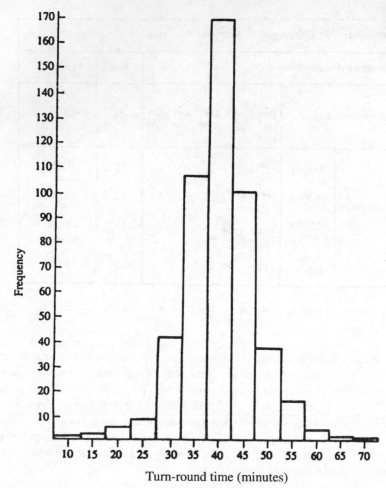

Figure 8.3   *Frequency distribution for truck turnround times (histogram)*

Stratification

Stratification is simply dividing a set of data into meaningful groups. It can be used to great effect in combination with other techniques, including histograms and scatter diagrams. If, for example, three shift teams are responsible for the output described by the histogram (a) in Figure 8.5, 'stratifying' the data into the shift groups might produce histograms (b), (c) and (d), and indicate process adjustments that were taking place at shift change overs.

Pareto analysis

If the symptoms or causes of defective output or some other 'effect' are identified and recorded, it will be possible to determine what percentage can be attributed to any cause, and the probable results will be that the bulk (typically 80 per cent) of the errors, waste, or 'effects', derive from a few of the causes (typically 20 per cent). For example,

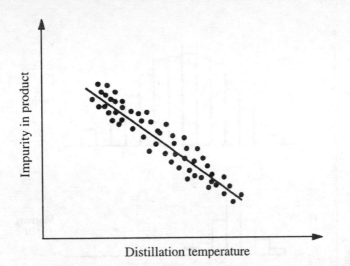

Figure 8.4   *Scatter diagram showing a negative correlation between two variables*

Figure 8.6 shows a *ranked frequency distribution* of incidents in the distribution of a certain product. To improve the performance of the distribution process, therefore, the major incidents (broken bags/drums, truck scheduling, temperature problems) should be tackled first. An analysis of data to identify the major problems is known as *Pareto analysis*, after the Italian economist who realised that approx 90 per cent of the wealth in his country was owned by approx 10 per cent of the people. Without an analysis of this sort, it is far too easy to devote resources to addressing one symptom only because its cause seems immediately apparent.

Cause and effect analysis and brainstorming

A useful way of mapping the inputs that affect quality is the *cause and effect diagram*, also known as the Ishikawa diagram (after its originator) or the fishbone diagram (after its appearance, Figure 8.7). The effect or incident being investigated is shown at the end of a horizontal arrow. Potential causes are then shown as labelled arrows entering the main cause arrow. Each arrow may have other arrows entering it as the principal factors or causes are reduced to their sub-causes, and sub-sub-causes by *brainstorming*.

Brainstorming is a technique used to generate a large number of ideas quickly, and may be used in a variety of situations. Each member of a group, in turn, may be invited to put forward ideas concerning a problem under consideration. Wild ideas are safe to offer, as criticism or ridicule is not permitted during a brainstorming session. The people taking part do so with equal status to ensure this. The main objective is to create an atmosphere of enthusiasm and originality. All ideas offered are recorded for subsequent analysis. The process is continued until all the conceivable causes have been included. The proportion of non-conforming output attributable to each cause, for example, is then measured or estimated, and a simple Pareto analysis identifies the causes that are most worth investigating.

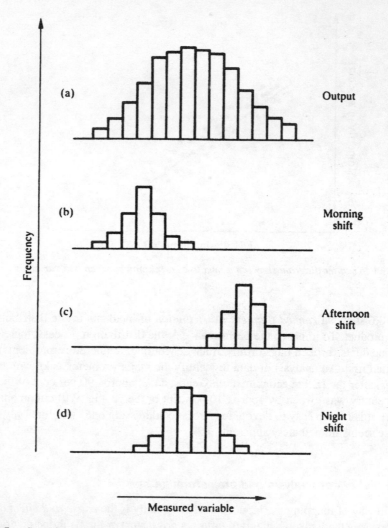

Figure 8.5   *Stratification of data into shift teams*

A useful variant on the technique is negative brainstorming and cause/effect analysis. Here the group brainstorms all the things that would need to be done to ensure a negative outcome. For example, in the implementation of TQM, it might be useful for the senior management team to brainstorm what would be needed to make sure TQM *was not* implemented. Having identified in this way the potential road blocks, it is easier to dismantle them.

CEDAC

A variation on the cause and effect approach, which was developed at Sumitomo Electric and now is claimed to be used by major Japanese corporations across the world, is the cause and effect diagram with addition of cards (CEDAC).

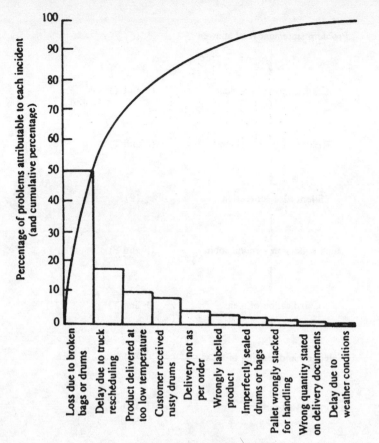

Figure 8.6   *Incidents in the distribution of a chemical product*

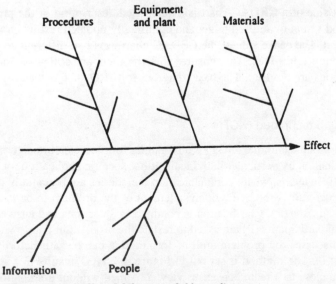

Figure 8.7   *The cause-and-effect, Ishikawa or fishbone diagram*

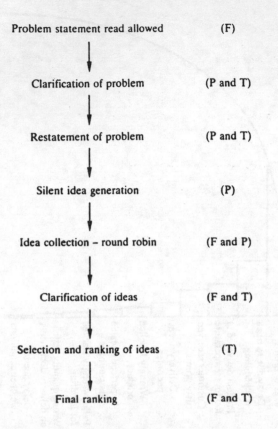

Problem statement read allowed　　　　(F)

Clarification of problem　　　　(P and T)

Restatement of problem　　　　(P and T)

Silent idea generation　　　　(P)

Idea collection – round robin　　　　(F and P)

Clarification of ideas　　　　(F and T)

Selection and ranking of ideas　　　　(T)

Final ranking　　　　(F and T)

Figure 8.8　*Nominal group technique (NGT)*

The effect side of a CEDAC chart is a quantified description of the problem, with an agreed and visual quantified target and continually updated results on the progress of achieving it. The cause side of the CEDAC chart uses two different coloured cards for writing facts and ideas. This ensures that the facts are collected and organised before solutions are devised. The basic diagram for CEDAC has the classic fishbone appearance.

Nominal group technique (NGT)

The nominal group technique (NGT) is a particular form of team brainstorming used to prevent domination by particular individuals. It has specific application for multi-level, multi-disciplined teams, where communication boundaries are potentially problematic.

In NGT a carefully prepared written statement of the problem to be tackled is read out by the facilitator (F). Clarification is obtained by questions and answers, and then the individual participants (P) are asked to restate the problem in their own words. The group then discusses the problem until its formulation can be satisfactorily expressed by the team (T). The method is set out in Figure 8.8. NGT results in a set of ranked ideas that are close to a team consensus view, obtained without domination by one or two individuals.

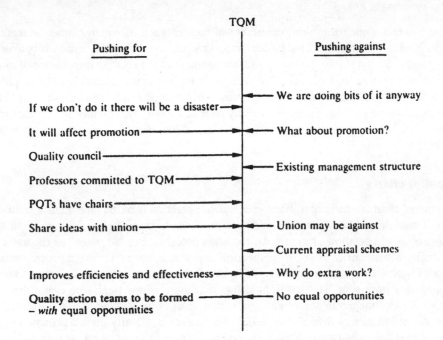

Figure 8.9   *Force-field analysis*

Even greater discipline may be brought to brainstorming by the use of 'Soft systems methodology (SSM)', developed by Peter Checkland.[3] The component stages of SSM are gaining a 'rich understanding' through 'finding out', input/output diagrams, root definition (which includes the so-called CATWOE analysis: customers, 'actors', transformations, 'world-view', owners, environment), conceptualisation, comparison, and recommendation.

Force field analysis

Force field analysis is a technique used to identify the forces that either obstruct or help a change that needs to be made. It is similar to negative brainstorming-cause/effect analysis and helps to plan how to overcome the barriers to change or improvement. It may also provide a measure of the difficulty in achieving the change.

The process begins with a team describing the desired change or improvement, and defining the objectives or solution. Having prepared the basic force field diagram, it identifies the favourable/positive/driving forces and the unfavourable/negative/restraining forces, by brainstorming. These forces are placed in opposition on the diagram and, if possible, rated for their potential influence on the ease of implementation. The results are evaluated. Then comes the preparation of an action plan to overcome some of the restraining forces, and increase the driving forces. Figure 8.9 shows a force field analysis produced by a senior management team considering the implementation of TQM in its organisation.

The emphasis curve

This is a technique for ranking a number of factors, each of which cannot be readily quantified in terms of cost, frequency or occurrence, etc, in priority order. It is almost impossible for the human brain to make a judgement of the relative importance of more than three or four non-quantifiable factors. It is, however, relatively easy to judge which is the most important of two factors, using some predetermined criteria. The emphasis curve technique uses this fact by comparing only two factors at any one time. The procedural steps for using the emphasis curve chart (matrix) are given by Oakland (1993).[4]

Control charts

A control chart is a form of traffic signal whose operation is based on evidence from the small samples taken at random during a process. A green light is given when the process should be allowed to run. All too often processes are 'adjusted' on the basis of a single measurement, check or inspection, a practice that can make a process much more variable than it is already. The equivalent of an amber light appears when trouble is possibly imminent. The red light shows that there is practically no doubt that the process has changed in some way and that it must be investigated and corrected to prevent production of defective material or information. Clearly, such a scheme can be introduced only when the process is 'in control'. Since samples taken are usually small, there are risks of errors, but these are small, calculated risks and not blind ones. The risk calculations are based on various frequency distributions.

These charts should be made easy to understand and interpret and they can become, with experience, sensitive diagnostic tools to be used by operating staff and first-line supervision to prevent errors or defective output being produced. Time and effort spent to explain the working of the charts to all concerned are never wasted.

The most frequently used control charts are simple run charts, where the data is plotted on a graph against time or sample number. There are different types of control charts for variables and attribute data: for variables mean ($\bar{x}$) and range (R) charts are used together; number defective or np charts and proportion defective or p charts are the most common ones for attributes. Other charts found in use are moving average and range charts, number of defects (c and u) charts, and cumulative sum (cusum) charts. The latter offer very powerful management tools for the detection of trends or changes in attributes and variable data.

The cusum chart is a graph that takes a little longer to draw than the conventional control chart, but gives a lot more information. It is particularly useful for plotting the evolution of processes, because it presents data in a way that enables the eye to separate true changes from a background of random variation. Cusum charts can detect small changes in data very quickly, and may be used for the control of variables and attributes. In essence, a reference or 'target value' is subtracted from each successive sample observation, and the result accumulated. Values of this cumulative sum are plotted, and 'trend lines' may be drawn on the resulting graphs. If they are approximately horizontal, the value of the variable is about the same as the target value. A downward slope shows a value less than the target, and an upward slope a value greater. The technique is very useful, for example, in comparing sales forecast with actual sales figures.

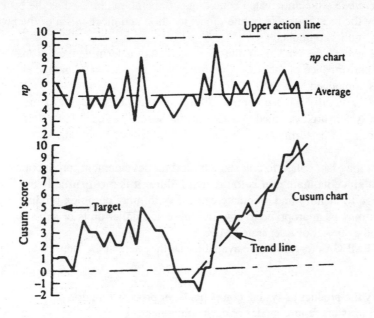

Figure 8.10 *Comparison of cusum and np charts for the same data*

Figure 8.10 shows a comparison of an ordinary run chart and a cusum chart that have been plotted from the same data – errors in samples of 100 invoices. The change, which is immediately obvious on the cusum chart, is difficult to detect on the conventional control chart.

The range of type and use of control charts is now very wide, and within the present text it is not possible to indicate more than the basic principles underlying such charts.[5]

8.3 Failure mode, effect and criticality analysis (FMECA)

It is possible to analyse products, services and processes to determine possible modes of failure and their effects on the performance of the product or operation of the process or service system. Failure mode and effect analysis (FMEA) is the study of potential failures to determine their effects. If the results of an FMEA are ranked in order of seriousness, then the word CRITICALITY is added to give FMECA. The primary objective of a FMECA is to determine the features of product design, production or operation and distribution that are critical to the various modes of failure, in order to reduce failure. It uses all the available experience and expertise, from marketing, design, technology, purchasing, production/operation, distribution, service, etc, to identify the importance levels or criticality of potential problems and stimulate action to reduce these levels. FMECA should be a major consideration at the design stage of a product or service (see Chapter 3).

The elements of a complete FMECA are:

- *Failure mode* – the anticipated conditions of operation are used as the background to study the most probable failure mode, location and mechanism of the product or system and its components.
- *Failure effect* – the potential failures are studied to determine their probable effects on the performance of the whole product, process, or service, and the effects of the various components on each other.
- *Failure criticality* – the potential failures on the various parts of the product or service system are examined to determine the severity of each failure effect in terms of lowering of performance, safety hazard, total loss of function, etc.

FMECA may be applied to any stage of design, development, production/operation or use, but since its main aim is to prevent failure, it is most suitably applied at the design stage to identify and eliminate causes. With more complex product or service systems, it may be appropriate to consider these as smaller units or sub-systems, each one being the subject of a separate FMECA.

Special FMECA pro-formas are available and they set out the steps of the analysis as follows:

1 Identify the product or system components, or process function.
2 List all possible failure modes of each component.
3 Set down the effects that each mode of failure would have on the function of the product or system.
4 List all the possible causes of each failure mode.
5 Assess numerically the failure modes on a scale from 1 to 10. Experience and reliability data should be used, together with judgement, to determine the values, on a scale 1–10 for:
   P the probability of each failure mode occurring (1 = low, 10 = high).
   S the seriousness or criticality of the failure (1 = low, 10 = high).
   D the difficulty of detecting the failure before the product or service is used by the consumer (1 = easy, 10 = very difficult). See Table 8.2.
6 Calculate the product of the ratings, C = P x S x D, known as the criticality index or risk priority number (RPN) for each failure mode. This indicates the relative priority of each mode in the failure prevention activities.
7 Indicate briefly the corrective action required and, if possible, which department or person is responsible and the expected completion date.

Table 8.2   *Probability and seriousness of failure and difficulty of detection*

Value	1	2	3	4	5	6	7	8	9	10
P	low chance of occurrence ——————————————— almost certain to occur									
S	not serious, minor nuisance ——————————————— total failure, safety hazard									
D	easily detected ——————————————————— unlikely to be detected									

When the criticality index has been calculated, the failures may be ranked accordingly. It is usually advisable, therefore, to determine the value of C for each failure mode before completing the last columns. In this way the action required against each item can be judged in the light of the ranked severity and the resources available.

Moments of truth

(MoT) is a concept that has much in common with FMEA. The idea was created by Jan Carlzon,[6] CEO of Scandinavian Airlines (SAS) and was made popular by Albrecht and Zemke.[7] An MoT is the moment in time when a customer first comes into contact with the people, systems, procedures, or products of an organisation, which leads to the customer making a judgement about the quality of the organisation's services or products.

In MoT analysis the points of potential dissatisfaction are identified proactively, beginning with the assembly of process flow chart type diagrams. Every small step taken by a customer in his/her dealings with the organisation's people, products, or services is recorded. It may be difficult or impossible to identify all the MoTs, but the systematic approach should lead to a minimalisation of the number and severity of unexpected failures, and this provides the link with FMEA.

8.4 Statistical process control (SPC)

The responsibility for quality in any transformation process must lie with the operators of that process. To fulfil this responsibility, however, people must be provided with the tools necessary to:

- Know whether the process is capable of meeting the requirements.
- Know whether the process is meeting the requirements at any point in time.
- Make correct adjustment to the process or its inputs when it is not meeting the requirements.

The techniques of statistical process control (SPC) will greatly assist in these stages. To begin to monitor and analyse any process, it is necessary first of all to identify what the process is, and what the inputs and outputs are. Many processes are easily understood and relate to known procedures, eg drilling a hole, compressing tablets, filling cans with paint, polymerising a chemical using catalysts. Others are less easily identifiable, eg servicing a customer, delivering a lecture, storing a product in a warehouse, inputting to a computer. In many situations it can be extremely difficult to define the process. For example, if the process is inputting data into a computer terminal, it is vital to know if the scope of the process includes obtaining and refining the data, as well as inputting. Process definition is so important because the inputs and outputs change with the scope of the process.

Once the process is specified, the inputs and suppliers, outputs and customers can also be defined, together with the requirements at each of the interfaces. The most difficult areas in which to do this are in non-manufacturing organisations or parts of organisations, but careful use of the questioning method, introduced in Chapter 1, should release the necessary information. Examples of outputs in non-manufacturing include training courses or programmes, typed letters, statements of intent (following a decision process), invoices, share certificates, deliveries of consignments, reports, serviced motor cars, purchase orders, wage slips, forecasts, material requirements plans, legal contracts, design change documents, clean offices, recruited trainees, and advertisements. The list is endless. Some processes may produce primary and secondary outputs, such as a telephone call answered *and* a message delivered.

If the requirements are not clarified or quantified, they are often assumed or estimated. Even if this does not lead to direct complaints, it will lead to waste – lost time, confusion – and perhaps lost customers. It is salutary for some suppliers of internal customers to realise that the latter can sometimes find new suppliers if their true requirements are not properly identified and/or repeatedly not met.

Inputs to processes include:

1 Equipment, tools, or plant required.
2 Materials – including paper.
3 Information – including the specification for the outputs.
4 Methods or procedures – including instructions.
5 People (and the inputs they provide, such as skills, training, knowledge, etc).
6 Records.

Again this is not an exhaustive list.

Prevention of failure in any transformation is possible only if the process definition, flow, inputs, and outputs are properly documented and agreed. The documentation of procedures will allow reliable data about the process itself to be collected, analysis to be performed, and action to be taken to improve the process and prevent failure or non-conformance with the requirements. The target in the operation of any process is the total avoidance of failure. If the idea of no failures or error free work is not adopted, at least as a target, then it certainly will never be achieved.

All processes can be monitored and brought 'under control' by gathering and using data – to measure the performance of the process and provide the feedback required for corrective action, where necessary. Statistical process control (SPC) methods, backed by management commitment and good organisation, provide objective means of *controlling* quality in any transformation process, whether used in the manufacture of artefacts, the provision of services, or the transfer of information.

SPC is not only a tool kit, it is a strategy for reducing variability, the cause of most quality problems: variation in products, in times of deliveries, in ways of doing things, in materials, in people's attitudes, in equipment and its use, in maintenance practices, in everything. Control by itself is not sufficient. Total quality management requires that the processes should be improved continually by reducing variability. This is brought about by studying all aspects of the process, using the basic question: 'Could we do this job more consistently and on target?' The answer drives the search for improvements. This significant feature of SPC means that it is not constrained to measuring conformance, and that it is intended to lead to action on processes that are operating within the 'specification' to minimise variability.

Process control is essential, and SPC forms a vital part of the TQM strategy. Incapable and inconsistent processes render the best design impotent and make supplier quality assurance irrelevant. Whatever process is being operated, it must be reliable and consistent. SPC can be used to achieve this objective.

In the application of SPC there is often an emphasis on techniques rather than on the implied wider managerial strategies. It is worth repeating that SPC is not only about plotting charts on the walls of a plant or office, it must become part of the company-wide adoption of TQM and act as the focal point of never-ending improvement. Changing an organisation's environment into one in which SPC can operate properly may take several years rather than months. For many companies SPC will being a new approach, a new 'philosophy', but the importance of the statistical techniques should not be disguised. Simple presentation of data using diagrams, graphs, and charts should become the means of communication concerning the state of control of processes. It is on this understanding that improvements will be based.

The SPC system

A systematic study of any process through answering the questions:

Are we capable of doing the job correctly?
Do we continue to do the job correctly?
Have we done the job correctly?
Could we do the job more consistently and on target?[8]

provides knowledge of the *process capability* and the sources of non-conforming outputs. This information can then be fed back quickly to marketing, design, and the 'technology' functions. Knowledge of the current state of a process also enables a more balanced judgement of equipment, both with regard to the tasks within its capability and its rational utilisation.

Statistical process control procedures exist because there is variation in the characteristics of all material, articles, services, and people. The inherent variability in each transformation process causes the output from it to vary over a period of time. If this variability is considerable, it is impossible to predict the value of a characteristic of any single item or at any point in time. Using statistical methods, however, it is possible to take meagre knowledge of the output and turn it into meaningful statements that may then be used to describe the process itself. Hence, statistically based process control procedures are designed to divert attention from individual pieces of data and focus it on the process as a whole. SPC techniques may be used to measure and control the degree of variation of any purchased materials, services, processes, and products, and to compare this, if required, to previously agreed specifications. In essence, SPC techniques select a representative, simple, random sample from the 'population', which can be an input to or an output from a process. From an analysis of the sample it is possible to make decisions regarding the current performance of the process.

8.5   Quality improvement techniques in non-manufacturing

Organisations that embrace the TQM concepts should recognise the value of SPC techniques in areas such as sales, purchasing, invoicing, finance, distribution, training, and in the service sector generally. These are outside the traditional areas for SPC use, but SPC needs to be seen as an organisation-wide approach to reducing variation with the specific techniques integrated into a programme of change throughout. A Pareto analysis, a histogram, a flowchart, or a control chart is a vehicle for communication. Data are data and, whether the numbers represent defects or invoice errors, weights or delivery times, or the information relates to machine settings, process variables, prices, quantities, discounts, sales or supply points, is irrelevant – the techniques can always be used.

In the authors' experience, some of the most exciting applications of SPC have emerged from organisations and departments which, when first introduced to the methods, could see little relevance in them to their own activities. Following appropriate training, however, they have learned how to, for example:

- *Pareto analyse* errors on invoices to customers and industry injury data.
- *Brainstorm and cause and effect analyse* reasons for late payment and poor purchase invoice matching.
- *Histogram* defects in invoice matching and arrival of trucks at certain times during the day.
- *Control chart* the weekly demand of a product.

Distribution staff have used control charts to monitor the proportion of late deliveries, and Pareto analysis and force field analysis to look at complaints about the distribution system. Word processor operators have been seen using cause and effect analysis, NGT and histograms to represent errors in the output from their service. Moving average and cusum charts have immense potential for improving processes in the marketing area.

Those organisations that have made most progress in implementing continuous improvement have recognised at an early stage that SPC is for the whole organisation. Restricting it to traditional manufacturing or operational activities means that a window of opportunity for improvement has been closed. Applying the methods and techniques outside manufacturing will make it easier, not harder, to gain maximum benefit from an SPC programme.

Sales, marketing and customer-service are areas often resistant to SPC training on the basis that it is difficult to apply. Personnel in these vital functions need to be educated in SPC methods for two reasons:

1 They need to understand the way the manufacturing or service producing processes in their organisations work. This will enable them to have more meaningful dialogues with customers about the whole product/service/delivery system capability and control. It will also enable them to influence customers' thinking about specifications and create a competitive advantage from improving process capabilities.
2 They will be able to improve the marketing processes and activities. A significant part of the sales and marketing effort is clearly associated with building relationships, which are best built on facts (data) and not opinions. There are also opportunities to use SPC techniques directly in such areas as forecasting demand levels and market requirements, monitoring market penetration, marketing control, and product development, all of which must be viewed as processes.

SPC has considerable applications for non-manufacturing organisations, including universities! Data and information on patients in hospitals, students in universities, polytechnics, colleges and schools, people who pay (and do not pay) tax, draw social security benefit, shop at Safeway's or Myer's, are available in abundance. If the information were to be used in a systematic way, and all operations treated as processes, far better decisions could be made concerning past, present, and future performances of some service sectors.

Chapter highlights

A systematic approach

- Numbers and information will form the basis for understanding, decisions, and actions in never ending improvement.
- A set of simple tools is needed to interpret fully and derive maximum use from data. More sophisticated techniques may need to be employed occasionally.
- The effective use of the tools requires the commitment of the people who work on the processes. This in turn needs management support and the provision of training.

Some basic tools and techniques

- The basic tools and the questions answered are:

Process flow charting	– what is done?
Check/tally charts	– how often is it done?
Histograms	– what do variations look like?
Scatter diagrams	– what are the relationships between factors?
Stratification	– how is the data made up?
Pareto analysis	– which are the big problems?
Cause and effect analysis and brainstorming (also CEDAC and NGT)	– what causes the problem?
Force-field analysis	– what will obstruct or help the change or solution?
Emphasis curve	– which are the most important factors?
Control charts (including cusum)	– which variations to control and how?

Failure mode, effect and criticality analysis (FMECA)

- FMEA is the study of potential product, service, or process failures and their effects. When the results are ranked in order of criticality, the technique is called FMECA. Its aim is to reduce the probability of failure.
- The elements of a complete FMECA are to study failure mode, effect, and criticality. It may be applied at any stage of design, development, production/operation, or use.
- Moments of truth (MoT) is a similar concept to FMEA. It is the moment in time when a customer first comes into contact with an organisation, leading to a judgement about quality.

Statistical process control

- People operating a process must know whether it is capable of meeting the requirements, know whether it is actually doing so at any time, and make correct adjustments when it is not. SPC techniques will help here.
- Before using SPC, it is necessary to identify what the process is, what the inputs/outputs are, and how the suppliers and customers and their requirements are defined. The most difficult areas for this can be in non-manufacturing.
- All processes can be monitored and brought 'under control' by gathering and using data. SPC methods, with management commitment, provide objective means of controlling quality in any transformation process.
- SPC is not only a toolkit, it is a strategy for reducing variability, part of never ending improvement. This is achieved by answering the following questions:

  Are we capable of doing the job correctly?
  Do we continue to do the job correctly?
  Have we done the job correctly?
  Could we do the job more consistently and on target?

  This provides knowledge of process capability.

Quality improvement techniques in non-manufacturing

- SPC techniques have value in the service sector and in the non-manufacturing areas, such as marketing and sales, purchasing, invoicing, finance, distribution, training and personnel.

References

1 John Oakland, *Total Quality Management*, 2nd edition, Butterworth–Heinemann, Oxford, 1993.
2 John Oakland and Roy Followell, *Statistical Process Control*, 2nd edition, Butterworth–Heinemann, Oxford, 1990.
3 Peter Checkland, *Soft Systems Methodology in Action*, Wiley, 1990.
4 See Oakland and Followell, *op cit.*
5 *Ibid.*
6 Jan Carlzon, *Moments of Truth*, Harper & Row, 1987.
7 Albrecht, K and Zemke, R, *Service America! – doing business in the new economy*, Dow Jones–Irwin, Homewood, Ill (USA), 1985.
8 This system for process capability and control is based on Frank Price's very practical framework for thinking about quality in manufacturing:

   Can we make it OK?
   Are we making it OK?
   Have we made it OK?
   Could we make it better?

   which he presented in his excellent book *Right First Time*.

9

Some additional techniques for process improvement

9.1 Seven new tools for quality design

Seven new tools may be used as part of quality function deployment (see Chapter 3) to improve the innovation processes. These do not replace the basic systematic tools described previously in Chapter 8, neither are they extensions of these. The new tools are systems and documentation methods used to achieve success in design by identifying objectives and intermediate steps in the finest detail. The seven new tools are:

1 Affinity diagram.
2 Interrelationship digraph.
3 Tree diagram.

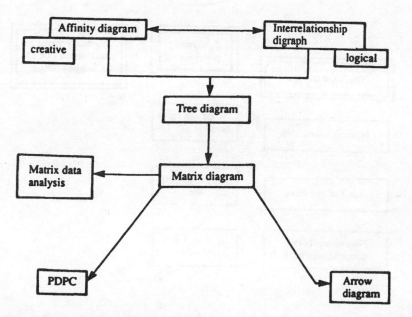

Figure 9.1 *The seven new tools of quality design*

4 Matrix diagram or quality table.
5 Matrix data analysis.
6 Process decision programme chart (PDPC).
7 Arrow diagram

   The tools are interrelated, as shown in Figure 9.1. The promotion and use of the tools by the QFD Team should obtain better designs in less time. They are summarised below and described in more detail in reference 1.

1   *Affinity diagram*

This is used to gather large amounts of language data (ideas, issues, opinions) and organises them into groupings based on the natural relationship between the items. In other words, it is a form of brainstorming. The affinity diagram is not recommended when a problem is simple or requires a very quick solution.

   The output of the exercise is a compilation of a maximum number of ideas under a limited number of major headings (see, for example, Figure 9.2). This data can then be used with other tools to define areas for attack. One of these tools is the interrelationship digraph.

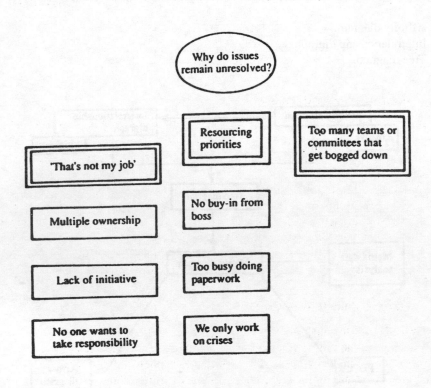

Figure 9.2   *Example of an affinity diagram*

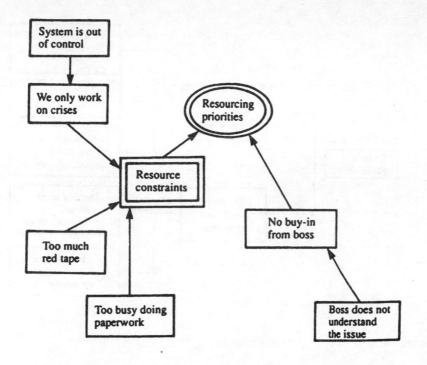

Figure 9.3 *Example of the interrelationship digraph*

2 Interrelationship digraph

This tool is designed to take a central idea, issue or problem, and map out the logical or sequential links among related factors. While this still requires a very creative process, the interrelationship digraph begins to draw the logical connections that surface in the affinity diagram.

The interrelationship digraph is adaptable to both specific operational issues and general organisational questions. For example, a classic use of this tool at Toyota focused on all the factors behind the establishment of a 'billboard system' as part of their JIT programme. On the other hand, it has also been used to deal with issues underlying the problem of getting top management support for TQM.

Figure 9.3 gives an example of a simple interrelationship digraph.

3 Systems flow/tree diagram

The systems flow/tree diagram (usually referred to as a tree diagram) is used to systematically map out the full range of activities that must be accomplished in order to reach a desired goal. It may also be used to identify all the factors contributing to a problem under consideration. Major factors identified by an interrelationship digraph can be used as inputs for a tree diagram. One of the strengths of this method is that it forces the user to examine the logical and chronological link between tasks. This assists in

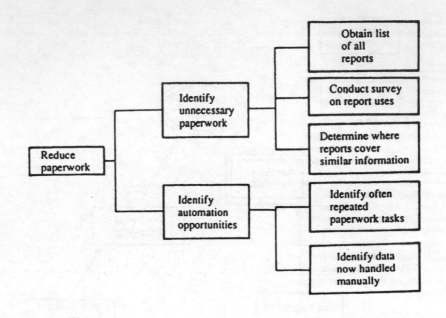

Figure 9.4   *An example of the tree diagram*

preventing a natural tendency to jump directly from goal or problem statement to solution (Ready … Fire … Aim!).

An example is shown in Figure 9.4.

4   *Matrix diagrams*

The matrix diagram is the heart of the seven new tools and the house of quality described in Chapter 3. The purpose of the matrix diagram is to outline the interrelationships and correlations between tasks, functions or characteristics, and to show their relative importance. There are many versions of the matrix diagram, but the most widely used is a simple L-shaped matrix known as the *quality table*.

Quality table

In a *quality table* customer demands (the whats) are analysed with respect to substitute quality characteristics (the hows), eg Figure 9.5. Correlations between the two are categorised as strong, moderate and possible. The customer demands shown on the left of the matrix are determined in co-operation with the customer. This effort requires a kind of a verbal 'ping-pong' with the customer to be truly effective: ask the customer what he wants, write it down, show it to him and ask him if that is what he meant, then revise and repeat the process as necessary. This should be done in a joint meeting with the customer, if at all possible. It is often of value to use a tree diagram to give structure to this effort.

Customer demands	Substitute quality characteristics: MFR	Substitute quality characteristics: Ash	Importance	Current	Best competitor	Plan	IR	SP	RQW
No film breaks	○ 17	▲ 6	4	4	4	4	1	○	5.6
High rates	◉ 23		3	3	4	4	1.3		4.6
Low gauge variability	◉ 37	▲ 7	4	3	4	4	1.3	○	7.3

◉ Strong correlation
○ Some correlation
▲ Possible correlation

IR Improvement ratio
SP Sales point
RQW Relative quality weight

Figure 9.5 *An example of the matrix diagram (quality table)*

The right side of the chart is often used to compare current performance to competitors' performance, company plan, and potential sales points with reference to the customer demands. Weights are given to these items to obtain a 'relative quality weight', which can be used to identify the key customer demands. The relative quality weight is then used with the correlations identified on the matrix to determine the key quality characteristics.

A modification that is added to create the house of quality table is a second matrix that explores the correlations between the quality characteristics. This is done so that errors caused by the manipulation of variables in a one-at-a-time fashion can be avoided. This also gives indications of where designed experiments would be of use in the design process. In the training required for use of this technique, several hours should be dedicated to a detailed explanation of the steps in the construction of a quality table, and the system to be used to compare numerically the various items.

T-shaped matrix diagram

The T-shaped matrix is nothing more than the combination of two L-shaped matrix diagrams. It is based on the premise that two separate sets of items are related to a third set. Therefore A items are somehow related to both B and C items.

5   *Matrix data analysis*

Matrix data analysis is used to take data displayed in a matrix diagram and arrange them so that they can be more easily viewed and show the strength of the relationship between variables. It is used most often in marketing and product research. The concept behind matrix data analysis is fairly simple, but its execution (including data gathering) can be complex.

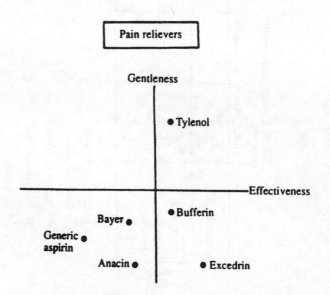

Figure 9.6   *An example of matrix data analysis*

A good idea of the uses and value of the construction of a chart for matrix data analysis may be shown in a simple example in which types of pain relievers are compared based on gentleness and effectiveness (Figure 9.6). This information could be used together with some type of demographic analysis to develop a marketing plan. Based on the information, advertising and product introduction could be effectively tailored for specific areas. New product development could also be carried out to attack specific niches in markets that would be profitable.

6   Process decision programme chart

A process decision programme chart (PDPC) is used to map out each event and contingency that can occur when progressing from a problem statement to its solution. The PDPC is used to anticipate the unexpected and plan for it. It includes plans for countermeasures on deviations. The PDPC is related to a failure mode and effect analysis and its structure is similar to that of a tree diagram. An example of the PDPC is shown in Figure 9.7.

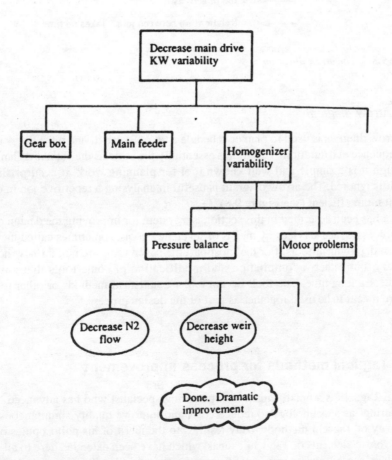

Figure 9.7   *Process decision programme chart*

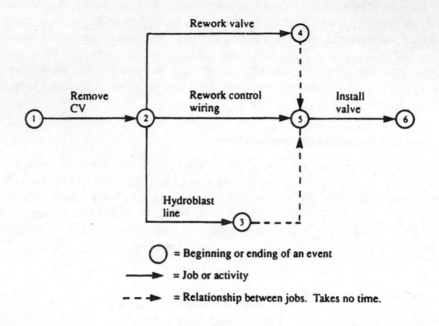

○ = Beginning or ending of an event

───▶ = Job or activity

─ ─ ▶ = Relationship between jobs. Takes no time.

Figure 9.8   *The arrow diagram*

7   Arrow diagram

The arrow diagram is used to plan or schedule a task. To use it, one must know the sub-task sequence and duration. This tool is essentially the same as the standard Gantt chart. Although it is a simple and well known tool for planning work, it is surprising how often it is ignored. The arrow diagram is useful in analysing a repetitive job in order to make it more efficient (see Figure 9.8).

What has been described in this section is a system for improving the design of products, processes, and services by means of seven new tools, sometimes called the quality function deployment tools. For the most part the seven tools are neither new nor revolutionary, but rather a compilation and modification of some tools that have been around for a long time. The tools do not replace statistical methods or other tools, but they are meant to be used together as part of the design process.

9.2   Taguchi methods for process improvement

Genichi Taguchi is a noted Japanese engineering specialist who has advanced 'quality engineering' as a technology to reduce costs and improve quality simultaneously. The popularity of Taguchi methods today testifies to the merit of his philosophies on quality. The basic elements of Taguchi's ideas, which have been extended here to all aspects of product, service and process quality, may be considered under four main headings.

1   Total loss function

An important aspect of the quality of a product or service is the total loss to society that it generates. Taguchi's definition of product quality as 'the loss imparted to society from the time a product is shipped', is rather strange, since the word *loss* denotes the very opposite of what is normally conveyed by using the word *quality*. The essence of his definition is that the smaller the loss generated by a product or service from the time it is transferred to the customer, the more desirable it is.

The main advantage of this idea is that it encourages a new way of thinking about investment in quality improvement projects, which become attractive when the resulting savings to customers are greater than the cost of improvements.

Taguchi claims with some justification that any variation about a target value for a product or process parameter causes loss to the customer. The loss may be some simple inconvenience, but it can represent actual cash losses, owing to rework or badly fitting parts, and it may well appear as loss of customer goodwill and eventually market share. The loss (or cost) increases exponentially as the parameter value moves away from the target, and is at a minimum when the product or service is at the target value.

2   Design of products, services and processes

In any product or service development three stages may be identified: product or service design, process design, and production or operations. Each of these overlapping stages has many steps, the output of one often being the input to others. The output/input transfer points between steps clearly affect the quality and cost of the final product or service. The complexity of many modern products and services demands that the crucial role of design be recognised. Indeed the performance of the quality products from the Japanese automotive, banking, camera, and machine tool industries can be traced to the robustness of their product and process designs.

The prevention of problems in using products or services under varying operating and environmental conditions must be built in at the design stage. Equally, the costs during production or operation are determined very much by the actual manufacturing or operating process. Controls, including SPC methods, added to processes to reduce imperfections at the operational stage are expensive, and the need for controls *and* the production of non-conformance can be reduced by correct initial designs of the process itself.

Taguchi distinguishes between *off-line* and *on-line* quality control methods, 'quality control' being used here in the very broad sense to include quality planning, analysis and improvement. Off-line QC uses technical aids in the *design* of products and processes, whereas on-line methods are technical aids for controlling quality and costs in the *production* of products or services. Too often the off-line QC methods focus on evaluation rather than improvement. The belief by some people (often based on experience!) that it is unwise to buy a new model of a motor car 'until the problems have been sorted out' testifies to the fact that insufficient attention is given to improvement at the product and process design stages. In other words, the bugs should be removed *before* not after product launch. This may be achieved in some organisations by replacing detailed quality and reliability evaluation methods with approximate estimates, and using the liberated resources to make improvements.

3   *Reduction of variation*

The objective of a continuous quality improvement programme is to reduce the variation of key products performance characteristics about their target values. The widespread practice of setting specifications in terms of simple upper and lower limits conveys the wrong idea that the customer is satisfied with all values inside the specification band, but is suddenly not satisfied when a value slips outside one of the limits. The practice of stating specifications as tolerance intervals only can lead manufacturers to produce and despatch goods whose parameters are just inside the specification band. Owing to the interdependence of many parameters of component parts and assemblies, this is likely to lead to quality problems.

The target value should be stated and specified as the ideal, with known variability about the mean. For those performance characteristics that cannot be measured on the continuous scale, the next best thing is an ordered categorical scale such as excellent, very good, good, fair, unsatisfactory, very poor, rather than the binary classification of 'good' or 'bad' that provides meagre information with which the variation reduction process can operate.

Taguchi has introduced a three-step approach to assigning nominal values and tolerances for product and process parameters:

(a) System design – the application of scientific engineering and technical knowledge to produce a basic functional prototype design. This requires a fundamental understanding of the needs of the customer *and* the production environment.
(b) Parameter design – the identification of the settings of product or process parameters that reduce the sensitivity of the designs to sources of variation. This requires a study of the whole process system design to achieve the most robust operational settings, in terms of tolerance to ranges of the input variables. This is similar to the experiments needed to identify the plant varieties that can tolerate variations in weather conditions, soil and handling. Manual processes that can tolerate the ranges of dimensions of the human body provide another example.
(c) Tolerance design – the determination of tolerances around the nominal settings identified by parameter design. This requires a trade-off between the customer's loss due to performance variation and the increase in production or operational costs.

4   *Statistically planned experiments*

Taguchi has pointed out that statistically planned experiments should be used to identify the settings of product and process parameters that will reduce variation in performance. He classifies the variables that affect the performance into two categories: design parameters and sources of 'noise'. As we have seen earlier, the nominal settings of the *design parameters* define the specification for the product or process. The *sources of noise* are all the variables that cause the performance characteristics to deviate from the target values. The *key* noise factors are those that represent the major sources of variability, and these should be identified and included in the experiments to design the parameters at which the effect of the noise factors on the performance is minimum. This is done by systematically varying the design parameter settings and comparing the effect of the noise factors for each experimental run.

Statistically planned experiments may be used to identify:

(a) The design parameters that have a large influence on the product or performance characteristic.
(b) The design parameters that have no influence on the performance characteristics (the tolerances of these parameters may be relaxed).
(c) The settings of design parameters at which the effect of the sources of noise on the performance characteristic is minimal.
(d) The settings of design parameters that will reduce cost without adversely affecting quality.[2]

Taguchi methods have stimulated a great deal of interest in the application of statistically planned experiments to product and process designs. The use of 'design of experiments' to improve industrial products and processes is not new – Tippett used these techniques in the textile industry more than 50 years ago. What Taguchi has done, however, is to acquaint us with the scope of these techniques in off-line quality control.

Taguchi's methods, like all others, should not be used in isolation, but be an integral part of continuous improvement.

9.3 Adding the tools to the TQM model

Having looked at some of the many tools and techniques of measurement and improvement, we see that the generic term 'tools' may be added, as the second hard management necessity, to the TQM model (Figure 9.9). The systems manage the processes, and

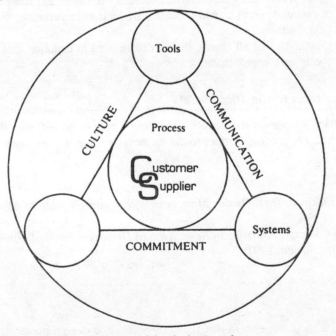

Figure 9.9   *Total quality management model – the basic tools*

the tools are used to progress further round the improvement cycle by creating better customer-supplier relationships, both externally and internally. They provide the means for analysis, correlation and prediction of what *action* to take on the systems.

Chapter highlights

Seven new tools for quality design

- Seven new tools may be used as part of quality function deployment (QFD, see Chapter 3) to improve the innovation processes. These are systems and documentation methods for identifying objectives and intermediate steps in the finest detail.
- The seven new tools are: affinity diagram, interrelationship digraph, tree diagram, matrix diagrams or quality table, matrix data analysis, process decision programme chart (PDPC), and arrow diagram.
- The tools are interrelated and their promotion and use should lead to better designs in less time. They work best when people from all parts of an organisation are using them. Some of the tools can be used in problem solving activities not related to design.

Taguchi methods for process improvement

- Genichi Taguchi has advanced 'quality engineering' as a technology to reduce costs and make improvements.
- Taguchi's approach may be classified under four headings; total loss function; design of products, services and processes; reduction in variation; and statistically planned experiments.
- Taguchi methods, like all others, should not be used in isolation, but as an integral part of continuous improvement.

Adding the tools to the TQM model

A second hard management necessity – the tools – may be added, with the systems, to the TQM model to progress further round the never-ending improvement cycle.

References

1   John S Oakland, *Total Quality Management*, 2nd edition, Butterworth–Heinemann, Oxford, 1993.
2   Roland Caulcutt, *Statistics in Research and Development*, 2nd edition, Chapman and Hall, London, 1991.

Discussion questions

1 (a) Using the expression: 'if you don't measure you can't improve', discuss this in the context of TQM. Why is measurement important?
   (b) What is the difference between measuring for results and measuring for process improvement? Using your knowledge of process management, where do you think measurement should take place and how should it be conducted?

2 Discuss the important features of a performance measurement system based on a TQM approach? Suggest an implementation strategy for a performance measurement system in a progressive company which is applying TQM principles to its business processes.

3 It is often said that 'you can't control what you can't measure and you can't manage what you can't control'. Measurement is, therefore, considered to be at the heart of managing business processes, activities and tasks. What do you understand by improvement-based performance measurement? Why is it important? Suggest a strategy of introducing TQM-based performance measurement for an organisation in the public sector.

4 Benchmarking is a new development in which most progressive organisations are interested. What do you understand by benchmarking? How does benchmarking link with performance measurement? Suggest a strategy for integrating benchmarking into a TQM programme.

5 (a) Some people would say that benchmarking is not different from competitor analysis and is a practice that organisations have always carried out. Do you agree with this? How would you define benchmarking and what are its key elements?
   (b) Suggest a benchmarking approach for a progressive small company that has no previous knowledge or experience of doing this.

6 (a) What are the major limitations of the 'Prevention–Appraisal–Failure (PAF)' costing model? Why would the process cost model be a better alternative?
   (b) Discuss the link between benchmarking and quality costing.
   (c) Suggest an implementation plan for benchmarking in a large company in the communications sector which is highly committed to TQM

7   A construction company is concerned about its record of completing projects on time. Considerable penalty costs are incurred if the company fails to meet the agreed contractual completion date. How would you investigate this problem and what methodology would you adopt?

8   English Aerospace is concerned about its poor delivery performance with the EA911. Considerable penalty costs are incurred if the company fails to meet agreed delivery dates.

As the company's TQM Manager you have been asked to investigate this problem using a systematic approach. Describe the methodology you would adopt.

9   The Marketing department of a large chemical company is reviewing its sales forecasting activities. Over the last three years the sales forecasts have been grossly inaccurate. As a result, a Quality Improvement Team has been formed to look at this problem. Give a systematic account of how the 'systematic tools' of TQM could be used in this situation.

10 It has been suggested by Deming and Ishikawa that statistical techniques can be used by staff at all levels within an organisation. Explain how such techniques can help:

(a) Senior managers to assess performance.
(b) Sales staff to demonstrate process capability to customers.
(c) Process teams to achieve quality improvement.

Part Four

TQM – The Organisational, Communication and Teamwork Requirements

Dust as we are, the immortal spirit grows
Like harmony in music; there is a dark
Inscrutable workmanship that reconciles
Discordant elements, makes them cling together
In one society.

William Wordsworth, 1770–1850

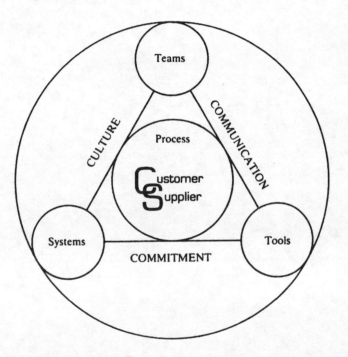

10

Organisation for quality

10.1 The quality function and the quality director or manager

In many organisations management systems are viewed in terms of the internal dynamics between marketing, design, sales, production/operations, distribution, accounting, etc. A change is required from this to a larger system that encompasses and integrates the business interests of customers and suppliers. Management needs to develop an in-depth understanding of these relationships and how they may be used to cement the partnership concept. The quality function should be the organisation's focal point in this respect, and should be equipped to gauge internal and external customers' expectations and degree of satisfaction. It should also identify quality deficiencies in all business functions, and promote improvements.

The role of the quality function is to make quality an inseparable aspect of every employee's performance and responsibility. The transition in many companies from quality departments with line functions will require careful planning, direction, and monitoring. Quality professionals have developed numerous techniques and skills, focused on product or service quality. In many cases there is a need to adapt these to broader applications. The first objectives for many 'quality managers' will be to gradually disengage themselves from line activities, which will then need to be dispersed throughout the appropriate operating departments. This should allow quality to evolve into a 'staff' department at a senior level, and to be concerned with the following throughout the organisation:

- Encouraging and facilitating quality improvement.
- Monitoring and evaluating the progress of quality improvement.
- Promoting the 'partnership' in quality, in relations with customers and suppliers.
- Planning, managing, auditing, and reviewing quality systems.
- Planning and providing quality training and counselling or consultancy.
- Giving advice to management on:

  (a) Establishment of quality systems and process control.
  (b) Relevant statutory/legislation requirements with respect to quality.
  (c) Quality improvement programmes necessary.
  (d) Inclusion of quality elements in all job instructions and procedures.

Quality directors and managers have an initial task, however, to help those who control the means to implement this concept – the leaders of industry and commerce –

to really believe that quality must become an integral part of all the organisation's operations.

The authors have a vision of quality as a strategic business management function that will help organisations to change their cultures. To make this vision a reality, quality professionals must expand the application of quality concepts and techniques to all business processes and functions, and develop new forms of providing assurance of quality at every supplier-customer interface. They will need to know the entire cycle of products or services, from concept to the *ultimate* end user. An example of this was observed in the case of a company manufacturing pharmaceutical seals, whose customer expressed concern about excess aluminium projecting below and round a particular type of seal. This was considered a cosmetic defect by the immediate customer, the Health Service, but a safety hazard by a blind patient – the *customer's customer*. The prevention of this 'curling' of excess metal meant changing practices at the mill that rolled the aluminium – at the *supplier's supplier*. Clearly, the quality professional dealing with this problem needed to understand the supplier's problems and the ultimate customer's needs, in order to judge whether the product was indeed capable of meeting the requirements.

The shift in 'philosophy' will require considerable staff education in many organisations. Not only must people in other functions acquire quality related skills, but quality personnel must change old attitudes and acquire new skills – replacing the inspection, calibration, specification-writing mentality and knowledge of defect prevention, wide ranging quality systems design and audit. Clearly, the challenge for many quality professionals is not so much making changes in their organisation as recognising the changes required in themselves. It is more than an overnight job to change the attitudes of an inspection police force into those of a consultative, team-oriented improvement force. This emphasis on prevention and improvement-based systems elevates the role of quality professionals from a technical one to that of general management. A narrow departmental view of quality is totally out of place in an organisation aspiring to TQM, and typical quality managers will need to widen their perspective and increase their knowledge to encompass all facets of the organisation.

To introduce the concepts of operator self-inspection required for TQM will require not only a determination to implement change but sensitivity and skills in industrial relations. This will depend very much of course on the climate within the organisation. Those whose management is truly concerned with co-operation and concerned for the people will engage strong employee support for the quality manager or director in his catalytic role in the quality improvement implementation process. Those with aggressive, confrontational management will create for the quality professional impossible difficulties in obtaining support from the 'rank and file'.

TQM appointments

Many organisations have realised the importance of the contribution a senior, qualified director of quality can make to the prevention strategy. Smaller organisations may well feel that the cost of employing a full-time quality manager is not justified, other than in certain very high risk areas. In these cases a member of the management team should be appointed to operate on a part-time basis, performing the quality management function in addition to his/her other duties. To obtain the best results from a quality director/manager, he/she should be given sufficient authority to take necessary action to secure the implementation of the organisation's quality policy, and must have the personality to be able to communicate the message to all employees, including staff, management and directors. Occasionally the quality director/manager may require some guidance and help on specific technical quality matters, and one of the major attributes required is the knowledge and wherewithal to acquire the necessary information and assistance.

In large organisations, then, it may be necessary to make several specific appointments or to assign details to certain managers. The following actions may be deemed to be necessary.

Assign a TQM director, manager or co-ordinator

This person will be responsible for the planning and implementation of TQM. He or she will be chosen first for project management ability rather than detailed knowledge of quality assurance matters. Depending on the size and complexity of the organisation, and its previous activities in quality management, the position may be either full or part-time, but it must report directly to the Chief Executive.

Appoint a quality management adviser

A professional expert on quality management will be required to advise on the 'technical' aspects of planning and implementing TQM. This is a consultancy role, and may be provided from within or without the organisation, full or part-time. This person needs to be a persuader, philosopher, teacher, adviser, facilitator, reporter and motivator. He or she must clearly understand the organisation, its processes and interfaces, be conversant with the key functional languages used in the business, and be comfortable operating at many organisational levels. On a more general level this person must fully understand and be an effective advocate and teacher of TQM, be flexible and become an efficient agent of change.

10.2 Councils, committees and teams

Devising and implementing total quality management for an organisation takes considerable time and ability. It must be given the status of a senior executive project. The creation of cost effective quality improvement is difficult, because of the need for full integration with the organisation's strategy, operating philosophy and management systems. It may require an extensive review and substantial revision of existing systems of management and ways of operating. Fundamental questions may have to be asked, such as 'Do the managers have the necessary authority, capability, and time to carry this through?'

Any review of existing management and operating systems will inevitably 'open many cans of worms' and uncover problems that have been successfully buried and smoothed over – perhaps for years. Authority must be given to those charged with following TQM through with actions that they consider necessary to achieve the goals. The commitment will be continually questioned and will be weakened, perhaps destroyed, by failure to delegate authoritatively.

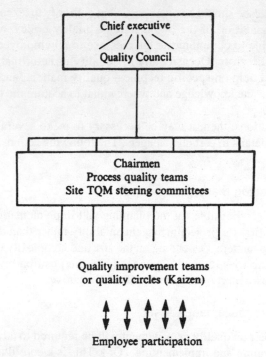

Figure 10.1   *Employee participation through the TQM structure*

The following steps are suggested in general terms. Clearly, different types of organisation will have need to make adjustments to the detail, but the component parts are the basic requirements.

A disciplined and systematic approach to continuous improvement may be established in a quality council (Figure 10.1). The council should meet at least monthly to review strategy, implementation progress, and improvement. It should be chaired by the Chief Executive, who must attend every meeting – only death or serious illness should prevent him/her being there. Clearly, postponement may be necessary occasionally, but the council should not carry on meeting without the Chief Executive present. The council members should include the top management team and the chairmen of any 'site' TQM steering committees or process quality teams, depending on the size of the organisation. The objectives of the council are to:

- Provide strategic direction on TQM for the organisation.
- Establish plans for TQM on each 'site'.
- Set up and review the process quality teams that will own the key or critical business processes.
- Review and revise quality plans for implementation.

The process quality teams (PQTs) and any site TQM steering committees should also meet monthly, shortly before the council meetings. Every senior manager should be a member of at least one PQT. This system provides the 'top-down' support for employee participation in process management and development, through either a quality improvement team or a quality circle programme. It also ensures that the commitment to TQM at the top is communicated effectively through the organisation.

The three-tier approach of quality council, process quality teams (PQTs) and quality improvement teams (QITs) allows the first to concentrate on quality strategy, rather than become a senior problem solving group. Progress is assured if the PQT chairmen are required to present a status report at each meeting.

The process quality teams or steering committees all control the QITs and have responsibility for:

- The selection of projects for the QITs.
- Providing an outline and scope for each project to give to the QITs.
- The appointment of team members and leaders.
- Monitoring and reviewing the progress and results from each QIT project.

As the focus of this work will be the selection of projects, some attention will need to be given to the sources of nominations. Projects may be suggested by:

(a) Council members representing their own departments, process quality teams, their suppliers or their customers, internal and external.
(b) Quality improvement teams.
(c) Quality circles (if in existence).
(d) Suppliers.
(e) Customers.

The PQT members must be given the responsibility and authority to represent their part of the organisation in the process. The members must also feel that they represent the team to the rest of the organisation. In this way the PQT will gain knowledge and respect and be seen to have the authority to act in the best interests of the organisation, with respect to their process.

10.3 Quality improvement teams

A quality improvement team (QIT) is a group of people with the appropriate knowledge, skills, and experience who are brought together specifically by management to tackle and solve a particular problem, usually on a project basis. They are cross functional and often multi-disciplinary.

The 'task force' has long been a part of the culture of many organisations at the 'technology' and management levels. But quality improvement teams go a step further; they expand the traditional definition of 'process' to cover the entire production or operating system. This includes paperwork, communication and other units, operating procedures, and the process equipment itself. By taking this broader view, the teams can address new problems.

The actual running of quality improvement teams calls several factors into play:

- Team selection and leadership.
- Team objectives.
- Team meetings.

- Team assignments.
- Team dynamics.
- Team results and reviews.

Team selection and leadership

The most important element of a QIT is its members. People with knowledge and experience relevant to solving the problem are clearly required; however, there should be a limit of five to ten members to keep the team small enough to be manageable but allow a good exchange of ideas. Membership should include appropriate people from groups outside the operational and technical areas directly 'responsible' for the problem, if their presence is relevant or essential. In the selection of team members it is often useful to start with just one or two people concerned directly with the problem. If they try to draw flowcharts (see Chapter 4) of the relevant processes, the requirement to include other people, in order to understand the process and complete the charts, will aid the team selection. This method will also ensure that all those who can make a significant contribution to the improvement process are represented.

The team leader has a primary responsibility for team management and maintenance, and his/her selection and training is crucial to success. The leader need not be the highest ranking person in the team, but must be concerned about accomplishing the team objectives (this is sometimes described as 'task concern') and the needs of the members (often termed 'people concern'). Weakness in either of these areas will lessen the effectiveness of the team in solving problems. Team leadership training should be directed at correcting deficiencies in these crucial aspects.

Team objectives

At the beginning of any QIT project and at the start of every meeting the objectives should be stated as clearly as possible by the leader. This can take a simple form: 'This meeting is to continue the discussion from last Tuesday on the provision of current price data from salesmen to invoice preparation, and to generate suggestions for improvement in its quality'. Project and/or meeting objectives enable the team members to focus thoughts and efforts on the aims, which may need to be restated if the team becomes distracted by other issues.

Team meetings

An agenda should be prepared by the leader and distributed to each team member before every meeting. It should include the following information:

- Meeting place, time and how long it will be.
- A list of members (and co-opted members) expected to attend.
- Any preparatory assignments for individual members or groups.
- Any supporting material to be discussed at the meeting.

Early in a project the leader should orient the team members in terms of the approach, methods, and techniques they will use to solve the problem. This may require a review of the:

1   Systematic approach (Chapter 8).
2   Procedures and rules for using some of the basic tools, eg brainstorming – no judge-
    ment of initial ideas.
3   Role of the team in the continuous improvement process.
4   Authority of the team.

A team secretary should be appointed to take the minutes of meetings and distribute them to members as soon as possible after each meeting. The minutes should not be formal, but reflect decisions and carry a clear statement of the action plans, together with assignments of tasks. They may be handwritten initially, copied and given to team members at the end of the meeting, to be followed later by a more formal document that will be seen by any member of staff interested in knowing the outcome of the meeting. In this way the minutes form an important part of the communication system, supplying information to other teams or people needing to know what is going on.

Team assignments

It is never possible to solve problems by meetings alone. What must come out of those meetings is a series of action plans that assign specific tasks to team members. This is the responsibility of the team leader. Agreement must be reached regarding the respon- sibilities for individual assignments, together with the time scale, and this must be made clear in the minutes. Task assignments must be decided while the team is together and not by separate individuals in after meeting discussions.

Team dynamics

In any team activity the interactions between the members are vital to success. If solu- tions to problems are to be found, the meetings and ensuing assignments should assist and harness the creative thinking process. This is easier said than done, because many people have either not learned or been encouraged to be innovative. The team leader clearly has a role here to:

*   Create a 'climate' for creativity.
*   Encourage all team members to speak out and contribute their own ideas or build on others.
*   Allow differing points of view and ideas to emerge.
*   Remove barriers to idea generation, eg incorrect preconceptions, which are usually destroyed by asking 'Why?'
*   Support all team members in their attempts to become creative.

In addition to the team leader's responsibilities, the members should:

(a) Prepare themselves well for meetings, by collecting appropriate data or information (*facts*) pertaining to a particular problem.
(b) Share ideas and opinions.
(c) Encourage other points of view.
(d) Listen 'openly' for alternative approaches to a problem or issue.

(e) Help the team determine the best solutions.
(f) Reserve judgement until all the arguments have been heard *and* fully understood.
(g) Accept individual responsibility for assignments and group responsibility for the efforts of the team.

Further details of teamworking are given in Chapter 11.

Teams results and reviews

A QIT approach to problem solving functions most effectively when the results of the projects are communicated and acted upon. Regular feedback to the teams, via their leaders, will assist them to focus on project objectives, and review progress.

Reviews also help to deal with certain problems that may arise in teamwork. For example, certain members may be concerned more with their own personal objectives than those of the team. This may result in some manipulation of the problem solving process to achieve different goals, resulting in the team splitting apart through self interest. If recognised, the review can correct this effect and demand greater openness and honesty.

A different type of problem is the failure of certain members to contribute and take their share of individual and group responsibility. Allowing other people to do their work results in an uneven distribution of effort, and leads to bitterness. The review should make sure that all members have assigned and specific tasks, and perhaps lead to the documentation of duties in the minutes. A team roster may even help.

A third area of difficulty, which may be improved by reviewing progress, is the ready-fire-aim syndrome of action before analysis. This often results from team leaders being too anxious to deal with a problem. A review should allow the problem to be redefined adequately and expose the real cause(s). This will release the trap the team may be in of doing something before they really know what should be done. The review will provide the opportunity to rehearse the steps in the systematic approach.

Record date	– all processes can and should be measured.
	– all measurements should be recorded.
Use data	– if data are recorded and not used they will be abused.
Analyse date systematically	– data analysis should be carried out by means of the basic tools (Chapter 8).
Act on the results	– recording and analysis of data without action leads to frustration.

10.4 Quality circles or Kaizen teams

Kaizen is a philosophy of continuous improvement of all the employees in an organisation, so that they perform their tasks a little better each day. It is a never-ending journey centred on the concept of starting anew each day with the principle that methods can always be improved. Using this approach, it is reported that Pratt and Whitney reduced reject rates on one process from 50 per cent to 4 per cent, and in 12 months eliminated overdue deliveries on a key sub-assembly.

Kaizen Teian is a Japanese system for generating and implementing employee ideas. Japanese suggestion schemes have helped companies to improve quality and productivity, and reduced prices to increase market share. They concentrate on participation and the rates of implementation, rather than on the 'quality' or value of the suggestion. The emphasis is on encouraging everyone to make improvements.

Kaizen Teian suggestions are usually small scale ones, in the worker's own area, and are easy and cheap to implement. Key points are that the rewards given are small, and implementation is rapid, which results in many small improvements that accumulate to massive total savings and improvements.

One of the most publicised aspects of the Japanese approach to quality has been quality circles or Kaizen teams. The quality circle may be defined as a group of workers doing similar work who meet:

- Voluntarily.
- Regularly.
- In normal working time.
- Under the leadership of their 'supervisor'.
- To identify, analyse, and solve work related problems.
- To recommend solutions to management.

Where possible quality circle members should implement the solutions themselves.

The quality circle concept first originated in Japan in the early 1960s, following a postwar reconstruction period during which the Japanese placed a great deal of emphasis on improving and perfecting their quality control techniques. As a direct result of work carried out to train foremen during that period, the first quality circles were conceived, and the first three circles registered with the Japanese Union of Scientists and Engineers (JUSE) in 1962. Since that time the growth rate has been phenomenal. The concept has spread to other Asian countries, the USA and Europe, and circles in many countries have become successful. Many others have failed.

It is very easy to regard quality circles as the magic ointment to be rubbed on the affected spot, and unfortunately many managers in the West have seen them as a panacea for all ills. There are no panaceas, and to place this concept into perspective, Juran, who has been an important influence in Japan's improvement in quality, has stated that quality circles represent only 5–10 per cent of the canvas of the Japanese success. The rest is concerned with understanding quality, its related costs and the organisation and techniques necessary for achieving customer satisfaction.

Given the right sort of commitment by top management, introduction, and environment in which to operate, quality circles can produce the 'shop floor' motivation to achieve quality performance at that level. Circles should develop out of an understanding and knowledge of quality on the part of senior management. They must not be introduced as a desperate attempt to do something about poor quality.

The structure of a quality circle organisation

The unique feature about quality circles or Kaizen teams is that people are asked to join and not told to do so. Consequently, it is difficult to be specific about the structure of such a concept. It is, however, possible to identify four elements in a circle organisation:

- Members.
- Leaders.
- Facilitators or co-ordinators.
- Management.

Members form the prime element of the programme. They will have been taught the basic problem solving and quality control techniques and, hence, possess the ability to identify and solve work related problems.

Leaders are usually the immediate supervisors or foremen of the members. They will have been trained to lead a circle and bear the responsibility of its success. A good leader, one who develops the abilities of the circle members, will benefit directly by receiving valuable assistance in tackling nagging problems.

Facilitators are the managers of the quality circle programmes. They, more than anyone else, will be responsible for the success of the concept, particularly within an organisation. The facilitator most co-ordinate the meetings, the training and energies of the leaders and members, and form the link between the circles and the rest of the organisation. Ideally the facilitator will be an innovative industrial teacher, capable of communicating with all levels and with all departments within the organisation.

Management support and commitment are necessary to quality circles or, like any other concept, they will not succeed. Management must retain its prerogatives, particularly regarding acceptance or non-acceptance of recommendations from circles, but the quickest way to kill a programme is to ignore a proposal arising from it. One of the most difficult facts for management to accept, and yet one forming the cornerstone of the quality circle philosophy, is that the real 'experts' on performing a task are those who do it day after day.

Training quality circles

The training of circle/Kaizen leaders and members is the foundation of all successful programmes. The whole basis of the training operation is that the ideas must be easy to take in and be put across in a way that facilitates understanding. Simplicity must be the key word, with emphasis being given to the basic techniques. Essentially there are eight segments of training:

1 Introduction to quality circles.
2 Brainstorming.
3 Data gathering and histograms.
4 Cause and effect analysis.
5 Pareto analysis.
6 Sampling.
7 Control charts.
8 Presentation techniques.

Managers should also be exposed to some training in the part they are required to play in the quality circle philosophy. A quality circle programme can only be effective if management believes in it and is supportive and, since changes in management style may be necessary, managers' training is essential.

Operation of quality circles/Kaizen teams

There are no formal rules governing the size of a quality circle/Kaizen team. Membership usually varies from three to fifteen people, with an average of seven to eight. It is worth remembering that, as the circle becomes larger than this, it becomes increasingly difficult for all members of the circle to participate.

Meetings must be held away from the work area, so that members are free from interruptions, and are mentally and physically at ease. The room should be arranged in a manner conductive to open discussion, and any situation that physically emphasises the leader's position should be avoided.

Meeting length and frequency are variable, but new circles meet for approximately one hour once per week. Thereafter, when training is complete, many circles continue to meet weekly; others extend the interval to 2 or 3 weeks. To a large extent the nature of the problems selected will determine the interval between meetings, but this should never extend to more than 1 month, otherwise members will lose interest and the circle will cease to function.

Great care is needed to ensure that every meeting is productive, no matter how long it lasts or how frequently it is held. Any of the following activities may take place during a circle meeting.

- Training – initial or refresher.
- Problem identification.
- Problem analysis.
- Preparation and recommendation for problem solution.
- Management presentations.
- Quality circle administration.

A quality circle usually selects a project to work on through discussion within the circle. The leader then advises management of this choice and, assuming that no objections are raised, the circle proceeds with the work. Other suggestions for projects come from management, quality assurance staff, the maintenance department, various staff personnel, and other circles.

It is sometimes necessary for quality circles to contact experts in a particular field, eg engineers, quality experts, safety officers, maintenance personnel. This communication should be strongly encouraged, and the normal company channels should be used to invite specialists to attend meetings and offer advice. The experts may be considered to be 'consultants', the quality circle retaining responsibility for solving the particular problem. The overriding purpose of quality circles or Kaizen teams is to provide the powerful motivation of allowing people to take some part in deciding their own actions and futures.

10.5 Departmental purpose analysis

'Quality is everyone's business' is an often quoted cliché, but 'Everything is everyone's business', and so quality often becomes nobody's business. The responsibility for quality begins with the determination of the customer's quality requirements and continues until the service or product is accepted by a satisfied customer. The department purpose analysis (DPA) technique, developed by IBM, helps to define the real purpose of each department, with the objective of improving performance and breaking down departmental barriers. It leads to an understanding and agreement on the key processes of each group. The department can then liaise with its immediate 'suppliers' and 'customers', often internally, to identify potential or actual problem areas and simultaneously carry out an analysis of what proportion of time is spent on the key activities. This begins the change from departmental to process management thinking.

Group discussions during the DPA process usually yield many good ideas for improvement, either eliminating wasteful activity or improving the quality of output from the department. Everyone becomes and should then remain aware of the prime purpose of the department, and the focus on efficiency and reducing waste usually carries through to all work activities. The manager of the department, who should run the exercise, must understand the DPA process and why it is necessary and important. He/she needs to be open minded towards change, and to encourage departmental staff to question whether all their activities add value to the product, service, or business. One of the greatest barriers to improvements through DPA is the 'but we've always done it that way' response.

The basic steps of DPA are:

1 Form the DPA group.
2 Brainstorm to list all the departmental tasks (see Chapter 8).
3 Agree which are the five main tasks.
4 Define the position and role of the departmental manager.
5 Review the main activities, and for each one identify the 'customer(s)' and 'supplier(s)'.
6 Consult the customer(s) and supplier(s) by means of a suitable questionnaire. This should be very similar to the list of questions suggested in Chapter 1 for interrogating any customer/supplier interface.
7 Review the customer/supplier survey results and brainstorm how improvements can be made.
8 Prioritise improvements to list those to be tackled first, and plan how.
9 Implement the improvement action plan, maintaining encouragement and support.
10 Review the progress made and repeat the DPA.

As with any new group activity, some successes are desirable early in the programme, if the department is to build confidence in its ability to make improvements and solve problems. For this reason DPA should confine itself, initially at least, to resolving issues that are within its control. It is unlikely, for example, that a sales team will be successful in getting a product redesigned in its first improvement project. Experience at IBM shows that, as confidence builds through continued management encouragement, the DPA groups will tackle increasingly difficult business processes and problems, with an increasing return of the investment in time.

Chapter highlights

The quality function and the quality director or manager

• The quality function should be the organisation's focal point of the integration of the business interests of customers and suppliers into the internal dynamics of the organisation.

- Its role is to encourage and facilitate quality improvement; monitor and evaluate progress; promote the quality chains; plan, manage, audit and review systems; plan and provide quality training, counselling and consultancy; and give advice to management.
- In larger organisations a quality director will contribute to the prevention strategy. Smaller organisations may appoint a member of the management team to this task on a part-time basis. An external TQM adviser is usually required.

Councils, committees and teams

- In devising and implementing TQM for an organisation, it may be useful to ask first if the managers have the necessary authority, capability and time to carry it through.
- A disciplined and systematic approach to continuous improvement may be established in a quality council (QC), whose members are the senior management team.
- Reporting to the QC are the process quality teams (PQTs) or any site steering committees, which in turn control the quality improvement teams (QITs) and quality circles.

Quality improvement teams

- A QIT is a group brought together by management to tackle a particular problem on a project basis. The running of QITs includes several team factors: selection and leadership, objectives, meetings, assignments, dynamics, results and reviews.

Quality circles or Kaizen teams

- Kaizen is a philosophy of small step continuous improvement, by all employees. In Kaizen teams the suggestions and rewards are small but the implementation is rapid.
- A quality circle or Kaizen team is a group of people who do similar work meeting voluntarily, regularly, in normal working time, to identify, analyse and solve work related problems, under the leadership of their supervisor. They make recommendations to management.

Departmental purpose analysis

- DPA helps to define the real purpose of each department, with the objective of improving performance and breaking down barriers. It leads to an understanding and agreement on the key processes of each group.
- The departmental manager runs the exercise and must understand DPA. The basic steps are form DPA group; list all departmental tasks; agree five main tasks; define position and role of manager; identify task customer(s) and supplier(s), and consult, review and brainstorm improvements; prioritise; implement plan; review progress and repeat DPA.

11

Culture change through teamwork for quality

11.1 The need for teamwork

The complexity of most of the processes that are operated in industry, commerce and the services places them beyond the control of any one individual. The only efficient way to tackle process improvement or problems is through the use of some form of teamwork. The use of the team approach to problem solving has many advantages over allowing individuals to work separately:

- A greater variety of complex problems may be tackled – those beyond the capability of any one individual or even one department – by the pooling of expertise and resources.
- Problems are exposed to a greater diversity of knowledge, skill, experience, and are solved more efficiently.
- The approach is more satisfying to team members, and boosts morale and ownership through participation in problem solving and decision making.
- Problems that cross departmental or functional boundaries can be dealt with more easily, and the potential/actual conflicts are more likely to be identified and solved.
- The recommendations are more likely to be implemented than individual suggestions, as the quality of decision making in *good teams*, is high.

Most of these factors rely on the premise that people are willing to support any effort in which they have taken part or helped to develop.

When properly managed and developed, teams improve the process of problem solving, producing results quickly and economically. Teamwork throughout any organisation is an essential component of the implementation of TQM, for it builds trust, improves communications and develops interdependence. Much of what has been taught previously in management has led to a culture in the West of independence, with little sharing of ideas and information. Knowledge is very much like organic manure – if it is spread around it will fertilise and encourage growth, if it is kept closed in, it will eventually fester and rot.

Teamwork devoted to quality improvement changes the independence to interdependence through improved communications, trust and the free exchange of ideas, knowledge, data and information (Figure 11.1). The use of the face-to-face interaction

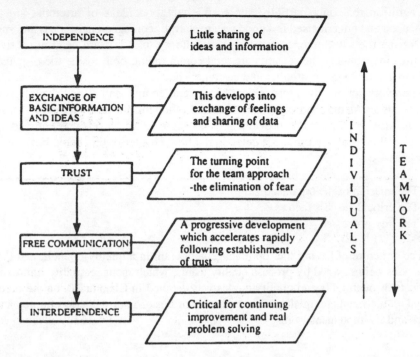

Figure 11.1   *Independence to interdependence through teamwork*

method of communication, with a common goal, develops over time the sense of dependence on each other. This forms a key part of any quality improvement process, and provides a methodology for employee recognition and participation, through active encouragement in group activities.

Teamwork provides an environment in which people can grow and use all the resources effectively and efficiently to make continuous improvements. As individuals grow, the organisation grows. It is worth pointing out, however, that employees will not be motivated towards continual improvement in the absence of:

- Commitment to quality for top management.
- The organisational quality 'climate'.
- A mechanism for enabling individual contributions to be effective.

All these are focused essentially at enabling people to feel, accept, and discharge responsibility. More than one organisation has made this part of their quality strategy – to 'empower people to act'. If one hears from employees comments such as 'We know this is not the best way to do this job, but if that is the way management want us to do it, that is the way we will do it', then it is clear that the expertise existing at the point of operation has not been harnessed and the people do not feel responsible for the outcome of their actions. Responsibility and accountability foster pride, job satisfaction, and better work.

Empowerment to act is very easy to express conceptually, but it requires real effort and commitment on the part of all managers and supervisors to put into practice.

Recognition that only partially successful but good ideas or attempts are to be applauded and not criticised is a good way to start. Encouragement of ideas and suggestions from the workforce, particularly through their part in team or group activities, requires investment. The rewards are total commitment, both inside the organisation and outside through the supplier and customer chains.

Teamwork for quality improvement has several components. It is driven by a strategy, needs a structure, and must be implemented thoughtfully and effectively. The strategy that drives the quality improvement teams at the various levels was outlined in Part 1, and will be dealt with in more detail in the final chapter of this book, but in essence it comprises:

- The mission of the organisation.
- The critical success factors.
- The key processes.

The structure of having the top management team in a quality council, and the key processes being owned by process quality teams, which manage quality improvement projects through QITs and quality circles was detailed in Chapter 10, on the organisational requirements for quality. The remainder of this chapter will concentrate on teamwork and its implementation.

11.2 Teamwork and action-centred leadership

Over the years there has been much academic work on the psychology of teams and on the leadership of teams. Three points on which all authors are in agreement are that teams develop a personality and culture of their own, respond to leadership, and are motivated according to criteria usually applied to individuals.

Key figures in the field of human relations, like Douglas McGregor (Theories X & Y), Abraham Maslow (Hierarchy of Needs) and Fred Hertzberg (Motivators and Hygiene Factors), all changed their opinions on group dynamics over time as they came to realise that groups are not the democratic entity that everyone would like them to be, but respond to individual, strong, well directed leadership, both from without and within the group, just like individuals.

Adair

During the 1960s John Adair, senior lecturer in Military History and the Leadership Training Adviser at the Military Academy, Sandhurst (UK) and later assistant director of the Industrial Society, developed what he called the action-centred leadership model, based on his experiences at Sandhurst, where he had the responsibility to ensure that results in the cadet training did not fall below a certain standard. He had observed that some instructors frequently achieved well above average results, owing to their own natural ability with groups and their enthusiasm. He developed this further into a team model, which is the basis of the approach of the authors and their colleagues to this subject.

In developing his model for teamwork and leadership, Adair brought out clearly that

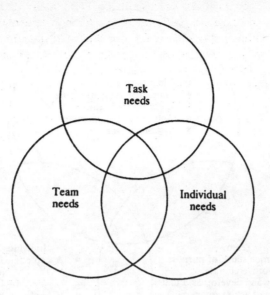

Figure 11.2 *Adair's model*

for any group or team, big or small, to respond to leadership, they need a clearly defined *task*, and the response and achievement of that task are interrelated to the needs of the *team* and the separate needs of the *individual members* of the team (Figure 11.2).

The value of the overlapping circles is that it emphasises the unity of leadership and the interdependence and multifunctional reaction to single decisions affecting any of the three areas.

Leadership tasks

Drawing upon the discipline of social psychology, Adair developed and applied to training the functional view of leadership. The essence of this he distilled into the three interrelated but distinctive requirements of a leader. These are to define and achieve the job or task, to build up and co-ordinate a team to do this, and to develop and satisfy the individuals within the team (Figure 11.3).

1 *Task needs*. The difference between a team and a random crowd is that a team has some common purpose, goal or objective, eg a football team. If a work team does not achieve the required results or meaningful results, it will become frustrated. Organisations have to make a profit, to provide a service, or even to survive. So anyone who manages others has to achieve results; in production, marketing, selling or whatever. Achieving objectives is a major criterion of success.

2 *Team needs*. To achieve these objectives, the group needs to be held together. People need to be working in a co-ordinated fashion in the same direction. Team work will ensure that the team's contribution is greater than the sum of its parts. Conflict within the team must be used effectively; arguments can lead to ideas or to tension and lack of co-operation.

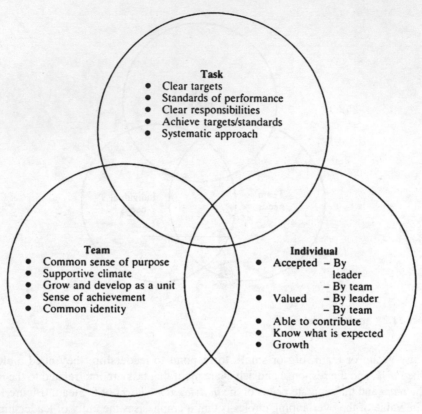

Figure 11.3 *The leadership needs*

3   *Individual needs*. Within working groups, individuals also have their own set of needs. They need to know what their responsibilities are, how they will be needed, how well they are performing. They need an opportunity to show their potential, take on responsibility and receive recognition for good work.

The task, team and individual functions for the leader are as follows:

(a)   *Task functions*   Defining the task.
                              Making a plan.
                              Allocating work and resources.
                              Controlling quality and tempo of work.
                              Checking performance against the plan.
                              Adjusting the plan.
(b)   *Team functions*   Setting standards.
                              Maintaining discipline.
                              Building team spirit.
                              Encouraging, motivating, giving a sense of purpose.
                              Appointing sub-leaders.
                              Ensuring communication within the group.
                              Training the group.

(c)   *Individual functions*   Attending to personal problems.
Praising individuals.
Giving status.
Recognising and using individual abilities.
Training the individual.

The team leader's or facilitator's task is to concentrate on the small central area where all three circles overlap. In a business that is introducing TQM this is the 'action to change' area, where the leaders are attempting to manage the change from *business as usual*, through total quality management, to *TQM equals business as usual*, using the cross-functional quality improvement teams at the strategic interface.

In the action area the facilitator's or leader's task is similar to the task outlined by John Adair. It is to try to satisfy all three areas of need by achieving the task, building the team, and satisfying individual needs. If a leader concentrates on the task, eg in going all out for production schedules, while neglecting the training, encouragement and motivation of the team and individuals, (s)he may do very well in the short term. Eventually, however, the team members will give less effort than they are capable of. Similarly, a leader who concentrates only on creating team spirit, while neglecting the task and the individuals, will not receive maximum contribution from the people. They may enjoy working in the team but they will lack the real sense of achievement that comes from accomplishing a task to the utmost of the collective ability.

So the leader/facilitator must try to achieve a balance by acting in all three areas of overlapping need. It is always wise to work out a list of required functions within the context of any given situation, based on a general agreement on the essentials. Here is Adair's original Sandhurst list, on which one's own adaptation may be based:

- *Planning*, eg seeking all available information.
  Defining group task, purpose or goal.
  Making a workable plan (in right decision-making framework).
- *Initiating*, eg briefing group on the aims and the plan.
  Explaining why aim or plan is necessary.
  Allocating tasks to group members.
  Setting group standards.
- *Controlling*, eg maintaining group standards.
  Influencing tempo.
  Ensuring all actions are taken towards objectives.
  Keeping discussion relevant.
  Prodding group to action/decision.
- *Supporting*, eg expressing acceptance of persons and their contribution.
  Encouraging group/individuals.
  Disciplining group/individuals.
  Creating team spirit.
  Relieving tension with humour.
  Reconciling disagreements or getting others to explore them.
- *Informing*, eg clarifying task and plan.
  Giving new information to the group.
  ie keeping them 'in the picture'.

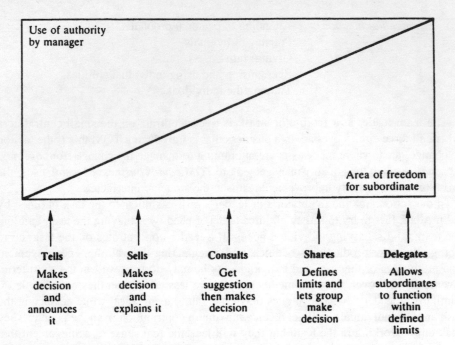

Figure 11.4   *Continuum of leadership behaviour*

Receiving information from group.

Summarising suggestions and ideas coherently.

• *Evaluating*, eg checking feasibility of an idea.

Testing the consequences of a proposed solution.

Evaluating group performance.

Helping the group to evaluate its own performance against standards.

Situational leadership

In dealing with the task, the team, and with any individual in the team, a style of leadership appropriate to the situation must be adopted. The teams and the individuals within them will, to some extent, start 'cold', but they will develop and grow in both strength and experience. The interface with the leader must also change with the change in the team, according to the Tannenbaum and Schmidt model (Figure 11.4).[1]

Initially a very directive approach may be appropriate, giving clear instructions to meet agreed goals. Gradually, as the teams become more experienced and have some success, the facilitating team leader will move through coaching and support to less directing and eventually a less supporting and less directive approach – as the more interdependent style permeates the whole organisation.

This equates to the modified Blanchard model[1] in Figure 11.5, where directive behaviour moves from high to low as people develop and are more easily empowered. When this is coupled with the appropriate level of supportive behaviour, a directing style of leadership can move through coaching and supporting to a delegating style. It must be stressed, however, that effective delegation is only possible with developed 'followers', who can be fully empowered.

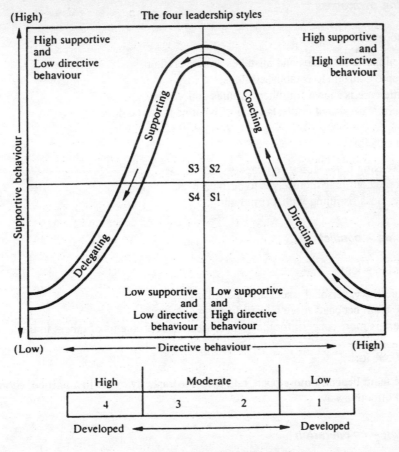

Figure 11.5 *Situational leadership – progressive empowerment through TQM*

One of the great mistakes in recent years has been the expectation by management that teams can be put together with virtually no training or development (S1 in Figure 11.5) and that they will perform as a mature team (S4). The Blanchard model emphasises that there is no quick and easy 'tunnel' from S1 to S4. The only route is the laborious climb through S2 and S3.

11.3 Stages of team development

Original work by Tuckman[1] suggested that when teams are put together, there are four main stages of team development, the so-called forming (awareness), storming (conflict), norming (co-operation), and performing (productivity). The characteristics of each stage and some key aspects to look out for in the early stages are given below:

Forming awareness

Characteristics:

- Feelings, weaknesses and mistakes are covered up.
- People conform to established lines.
- Little care is shown for others' values and views.
- There is no shared understanding of what needs to be done.

Watch out for:

- Increasing bureaucracy and paperwork.
- People confining themselves to defined jobs.
- The 'boss' is ruling with a firm hand.

Storming – conflict

Characteristics:

- More risky, personal issues are opened up.
- The team becomes more inward-looking.
- There is more concern for the values, views and problems of others in the team.

Watch out for:

- The team becomes more open, but lacks the capacity to act in a unified, economic, and effective way.

Norming – co-operation

Characteristics:

- Confidence and trust to look at how the team is operating.
- A more systematic and open approach, leading to a clearer and more methodical way of working.
- Greater valuing of people for their differences.
- Clarification of purpose and establishing of objectives.
- Systematic collection of information.
- Considering all options.
- Preparing detailed plans.
- Reviewing progress to make improvements.

Performing – productivity

Characteristics:

- Flexibility.
- Leadership decided by situations, not protocol.

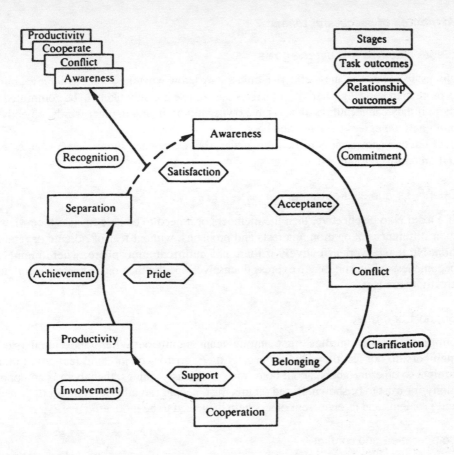

Figure 11.6 *Team stages and outcomes. (Derived from Kormanski and Mozenter, 1987)*

- Everyone's energies utilised.
- Basic principles and social aspects of the organisation's decisions considered.

The team stages, the task outcomes, and the relationship outcomes are shown together in Figure 11.6. This model, which has been modified from Kormanski and Mozenter,[2] may be used as a framework for the assessment of team performance. The issues to look for are:

1 How is leadership exercised in the team?
2 How is decision making accomplished?
3 How are team resources utilised?
4 How are new members integrated into the team?

Teams which go through these stages successfully should become effective teams and display the following attributes.

Attributes of successful teams

Clear objectives and agreed goals

No group of people can be effective unless they know what they want to achieve, but it is more than knowing what the objectives are. People are only likely to be committed to them if they can identify with and have ownership of them – in other words, objectives and goals are agreed by team members.

Often this agreement is difficult to achieve but experience shows that it is an essential prerequisite for the effective group.

Openness and confrontation

If a team is to be effective, then the members of it need to be able to state their views, their differences of opinion, interests and problems, without fear of ridicule or retaliation. No teams work effectively if there is a cutthroat atmosphere, where members become less willing or able to express themselves openly; then much energy, effort and creativity are lost.

Support and trust

Support naturally implies trust among team members. Where individual group members do not feel they have to protect their territory or job, and feel able to talk straight to other members, about both 'nice' and 'nasty' things, then there is an opportunity for trust to be shown. Based on this trust, people can talk freely about their fears and problems and receive from others help they need to be more effective.

Co-operation and conflict

When there is an atmosphere of trust, members are more ready to participate and are committed. Information is shared rather than hidden. Individuals listen to the ideas of others and build on them. People find ways of being more helpful to each other and the group generally. Co-operation causes high morale – individuals accept each other's strengths and weaknesses and contribute from their pool of knowledge of skill. All abilities, knowledge and experience are fully utilised by the group; individuals have no inhibitions about using other people's abilities to help solve their problems, which are shared.

Allied to this, conflicts are seen as a necessary and useful part of organisational life. The effective team works through issues of conflict and uses the results to help objectives. Conflict prevents teams from becoming complacent and lazy, and often generates new ideas.

Good decision-making

As mentioned earlier, objectives need to be clearly and completely understood by all members before good decision making can begin. In making decisions effective, teams develop the ability to collect information quickly then discuss the alternatives openly. They become committed to their decisions and ensure quick action.

Appropriate leadership

Effective teams have a leader whose responsibility it is to achieve results through the efforts of a number of people. Power and authority can be applied in many ways, and team members often differ on the style of leadership they prefer. Collectively, teams may come to different views of leadership but, whatever their view, the effective team usually sorts through the alternatives in an open and honest way.

Review of the team processes

Effective teams understand not only the group's character and its role in the organisation, but how it makes decisions, deals with conflicts, etc. The team process allows the team to learn from experience and consciously to improve teamwork. There are numerous ways of looking at team processes – use of an observer, by a team member giving feedback, or by the whole group discussing members' performance.

Sound inter-group relationships

No human being or group is an island; they need the help of others. An organisation will not achieve maximum benefit from a collection of quality improvement teams that are effective within themselves but fight among each other.

Individual development opportunities

Effective teams seek to pool the skills of individuals, and it necessarily follows that they pay attention to development of individual skills and try to provide opportunities for individuals to grow and learn, and of course have FUN.

Once again, these ideas are not new but are very applicable and useful in the management of teams for quality improvements, just as Newton's theories on gravity still apply!

11.4 Team roles and personality types

No one person has a monopoly of 'good' characteristics. Attempts to list the qualities of the ideal manager, for example, demonstrate why that paragon cannot exist. This is because many of the qualities are mutually exclusive, for example:

Highly intelligent	*v*	Not *too* clever
Forceful and dominant	*v*	Sensitive to people's feelings
Dynamic	*v*	Patient
Fluent communicator	*v*	Good listener
Decisive	*v*	Reflective

Although no individual can possess all these and more desirable qualities, a team often does.

The overwhelming majority of behavioural research has been concerned with the individual. Since the early 1980s, however, some very valuable work has at least been done on teams, including that of Dr Meredith Belbin. Through observation over many years, both in industry and in the world of management training, Belbin identified a set of eight 'roles' which, if all present in a team, give that team the best possible chance of success. Indeed the eight roles are the only ones available in a team, they are:

Co-ordinator (or chairman)
Shaper
Plant (ideas generator)
Monitor – evaluator
Implementor (or worker)
Resource investigator
Teamworker
Finisher

A preponderance of a few of these roles and the absence of some within the team are a pretty good guarantee of failure, whatever the intelligence, motivation, etc, of the individuals concerned.

A few general points are worth making about team roles:

- The term 'plant' is used because this type, if 'planted' in a bogged-down group, will get it going again.
- All roles have value, and are missed when not in a team. There are no 'stars' or 'extras'.
- In small teams people can and do assume more than one role.
- The roles divide generally into outward- and inward-looking groups:

Outward-looking	Inward-looking
Co-ordinator	Implementor
Plant	Monitor-evaluator
Resource investigator	Teamworker
Shaper	Finisher

- The team role for an individual is determined by the completion and analysis of Belbin's self-administered questionnaire.

Using team roles

Eight people are not required for a team, but people who are aware and capable of carrying out the roles should be present. A team will not perform so effectively if there is not a good match between the attributes of team members and their responsibilities, eg, if the co-ordinator is actually a shaper.

Most people play different roles to suit different situations. The natural or *main* role for an individual may be a shaper, but if there is already a strong shaper in the group, it may be advisable for him/her to develop a *secondary* role.

Some roles represent *active characteristics*, eg the shaper 'makes things happen' and the implementor 'converts plans into tasks'. Other roles are *passive descriptions* of personality, eg, the teamworker 'dislikes friction and confrontation' and the plant is 'forthright' and 'independent'. Groups need active members. They are not necessarily people who talk too much and dominate a meeting, but people who make a positive contribution to the proceedings.

Analysing existing groups and their performance or behaviour, using the team roles concept, can lead to improvement, eg:

- Under-achievement demands a good co-ordinator or finisher.
- Conflict within the group requires a teamworker or strong co-ordinator.
- Mediocre performance can be improved by a resource investigator, innovator or shaper.
- Error prone groups need a clever evaluator and an able organiser.

Stable organisations need a different mix of people to those operating in areas of rapid change. Different roles are more important in particular circumstances. For example, new groups need a strong shaper to get started, competitive situations demand an innovator with good ideas, and in areas of high risk a good evaluator may be needed. Teams should be analysed therefore, both in terms of what team roles members can play, and also in relation to what team skills are most needed.

The Belbin team roles concept has the merit of simplicity. The authors and their colleagues believe, however, that a more complete, understandable, and helpful approach is provided by the use of the personality type indicator described in the following section.

Understanding and valuing team members – the MBTI

A powerful aid to team development is the use of the Myers-Briggs Type Indicator (MBTI).[2] This is based on an individual's preferences on four scales for:

- Giving and receiving 'energy'.
- Gathering information.
- Making decisions.
- Handling the outer world.

Its aim is to help individuals understand and value themselves and others, in terms of their differences as well as their similarities. It is well researched and non-threatening when used appropriately.

The four MBTI preferences scales, which are based on Jung's theories of psychological types, represent two opposite preferences:

•	*Extroversion – Introversion*	– how we prefer to give/receive energy or focus our attention.
•	*Sensing – iNtuition*	– how we prefer to gather information.
•	*Thinking – Feeling*	– how we prefer to make decisions.
•	*Judgement – Perception*	– how we prefer to handle the outer world.

To understand what is meant by preferences, the analogy of left and right-handedness is useful. Most people have a preference to write with either their left or their right hand. When using the preferred hand, they tend not to think about it, it is done naturally. When writing with the other hand, however, it takes longer, needs careful concentration, seems more difficult, but with practice would no doubt become easier. Most people *can* write with and use both hands, but tend to prefer one over the other. This is similar to the MBTI psychological preferences: most people are able to use both preferences at different times, but will indicate a preference on each of the scales.

In all, there are eight possible preferences – E or I, S or N, T or F, J or P, ie two opposites for each of the four scales. An individual's *type* is the combination and interaction of the four preferences. It can be assessed initially by completion of a simple questionnaire. Hence, if each preference is represented by its letter, a person's type may be shown by a four letter code – there are sixteen in all. For example, ESTJ represents an *extrovert* (E) who prefers to gather information with *sensing* (S), prefers to make decisions by *thinking* (T) and has a *judging* (J) attitude towards the world, ie prefers to make decisions rather than continue to collect information. The person with opposite preferences on all four scales would be an INFP, an introvert who prefers intuition for perceiving, feelings or values for making decisions, and likes to maintain a perceiving attitude towards the outer world.

The questionnaire, its analysis and feedback must be administered by a qualified MBTI practitioner, who may also act as external facilitator to the team in its forming and storming stages.

Type and teamwork

With regard to teamwork, the preference types and their interpretation are extremely powerful. The *extrovert* prefers action and the outer world, whilst the *introvert* prefers ideas and the inner world.

Sensing-thinking types are interested in facts, analyse facts impersonally, and use a step-by-step process from cause to effect, premise to conclusion. The *sensing-feeling* combinations, however, are interested in facts, analyse facts personally, and are concerned about how things matter to themselves and others.

Intuition-thinking types are interested in possibilities, analyse possibilities impersonally, and have theoretical, technical, or executive abilities. On the other hand, the *intuition-feeling* combinations are interested in possibilities, analyse possibilities personally, and prefer new projects, new truths, things not yet apparent.

Judging types are decisive and planful, they live in orderly fashion, and like to regulate and control. *Perceivers*, on the other hand are flexible, live spontaneously, and understand and adapt readily.

As we have seen, an individual's type is the combination of four preferences on each of the scales. There are sixteen combinations of the preference scales and these may be displayed on a *type table* (Figure 11.7). If the individuals within a team are prepared to share with each other their MBTI preferences, this can dramatically increase understanding and frequently is of great assistance in team development and good team working. The similarities and differences in behaviour and personality can be identified. The assistance of a qualified MBTI practitioner is absolutely essential in the initial stages of this work.

The five 'A' stages for teamwork

For any of these models or theories to benefit a team, the individuals within it need to become *aware* of the theory, eg the MBTI. They then need to *accept* the principles as valid, *adopt* them for themselves in order to *adapt* their behaviour accordingly. This will lead to individual and team *action* (Figure 11.8).

In the early stages of team development particularly, the assistance of a skilled facilitator to aid progress through these stages is necessary. This is often neglected, causing failure in so many team initiatives. In such cases the net output turns out to be lots of nice warm feelings about 'how good that team workshop was a year ago', but the nagging reality that no action came out and nothing has really changed.

11.5 Implementing teamwork for quality improvement – the 'DRIVE' model

The authors and their colleagues have developed a model for a structured approach to problem solving in teams, the *DRIVE* model. The mnemonic provides landmarks to keep the team on track and in the right direction:

Define   – the problem. *Output*: written definition of the task and its success criteria.

Review   – the information. *Output*: presentation of known data and action plan for further data.

ISTJ	ISFJ	INFJ	INTJ
ISTP	ISFP	INFP	INTP
ESTP	ESFP	ENFP	ENTP
ESTJ	ESFJ	ENFJ	ENTJ

Figure 11.7 *MBTI type table form. Source: Isabel Briggs Myers, Introduction to Type*[2]

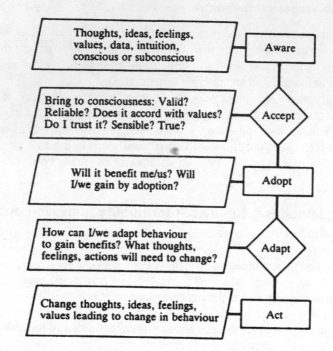

Figure 11.8 *The five 'A' stages for teamwork*

<u>I</u>nvestigate	– the problem. *Output*: documented proposals for improvement and action plans.
<u>V</u>erify	– the solution. *Output*: proposed improvements that meet success criteria.
<u>E</u>xecute	– the change. *Output*: task achieved and improved process documented.

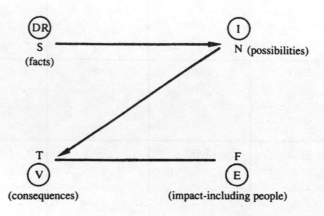

Figure 11.9 *The DRIVE model and MBTI-based problem-solving*

The DRIVE model fits well with the MBTI Z-shaped problem-solving approach. Figure 11.9 shows how the stages relate to the S-N-T-F path.

The various stages are discussed in detail in Oakland (1993).[3]

Steps in the introductions of teams

The idea of introducing problem solving groups, quality circles or quality improvement teams often makes its way into an organisation through the awareness of successful results in other organisations or companies. There is no fixed methodology for starting a teamwork programme, but there are certain key points that must be considered:

1   The concept should be presented to (or come from) management and supervision, and their commitment and support enlisted. It should be possible at this stage to engage the interest and support of potential team leaders.
2   Projects should be started slowly and on a small scale. Ideally a pilot scheme, run by the most enthusiastic candidates and in the most promising areas, should be launched. Early teething troubles, doubts and worries may then be identified and resolved.
3   Selected or volunteer team or circle leaders must be trained in all aspects of group leadership, and the appropriate techniques, and they should subsequently help train the team members in the techniques required in effective problem solving. The techniques of statistical process control (SPC) should be introduced, particularly brainstorming, cause and effect analysis, Pareto analysis and charting. These concepts lay the groundwork for analysing problems in a systematic fashion, and show that the majority of the problems are concentrated into a few areas.
4   Once the causes have been determined, a solution can be proposed. This solution may affect any of the components of the process: equipment, procedures, training, input requirements or output requirements. The proposed solution should be tested by the team or circle, particularly if procedures are affected.
5   If the test of a solution proves successful, full-scale implementation can then be carried out. In the case of procedures, full documentation of the solution and management approval should be obtained. The procedure can then be communicated to all personnel concerned. Full-scale changes in equipment and other processes should occur in the same manner. The team should monitor implementation of the solution, plotting the appropriate data until the criteria for solution are met.

With the initial problems declared solved, the circle or team may then tackle another problem, and another, or be disbanded and new teams formed. The record of successful solutions will motivate other teams within the organisation, and ideas should spread. As the number of teams in a company grows, new opportunities arise for stimulating interest. Some large companies organise in-house conferences of their quality improvement teams and quality circles, providing the opportunity for the publication of results and for recognition. Experience has shown that very significant improvements in areas such as energy reduction, productivity, and cost-effectiveness, in addition to quality, may be achieved by the project team approach.

One of the problems of the team approach to problem identification and solving is

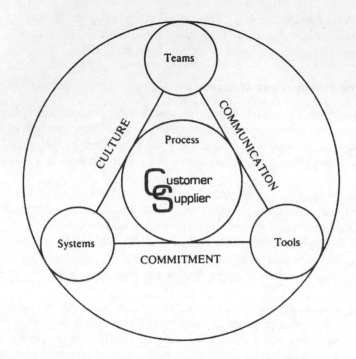

Figure 11.10 *Total quality management – teamwork added to complete the model*

that sometimes the teams are organised because it is the fashionable thing to do. They either exist on paper only, or the meetings are social gatherings where nothing is learned, no projects are initiated, and people do not grow. Another common problem is that the teams attempt to solve problems without first learning the necessary techniques: enthusiasm outruns ability. Teams have enormous potential for helping to solve an organisation's problems, but for them to be successful, they must follow a disciplined approach to problem solving, using proven techniques.

The team approach to problem solving works. It taps the skills and initiative of all personnel engaged in a process. This may mean a change in culture, which must be supported by management through its own activities and behaviour.

11.6 Adding the teams to the TQM model

In part 1 of this book the foundations for TQM were set down. The core of customer/supplier chains and, at every interface, a process were surrounded by the 'soft' outcomes of culture, communications, and commitment. In Parts 2 and 3 were added the hard management necessities of systems and tools. We are now ready to complete the model with the necessity of teams – the councils, the PQTs, QITs, quality circles, DPA groups, etc, which work on the processes – using the tools – to bring about continuous improvements in the systems that manage them. (Figure 11.10).

The authors are grateful for the significant contribution to this chapter made by their colleagues in O&F Quality Management Consultants Ltd, Stephen Mathews, Development Director and John Glover, Senior Consultant.

Chapter highlights

The need for teamwork

- The only efficient way to tackle process improvement or complex problems is through teamwork. The team approach allows individuals and organisations to grow.
- Employees will not engage continual improvement without commitment from the top, a quality 'climate', and an effective mechanism for capturing individual contributions.
- Teamwork for quality improvement is driven by a strategy, needs a structure, and must be implemented thoughtfully and effectively.

Teamwork and action-centred leadership

- Early work in the field of human relations by McGregor, Maslow, and Hertzberg was useful to John Adair in the development of his model for teamwork and action-centred leadership.
- Adair's model addresses the needs of the task, the team, and the individuals in the team, in the form of three overlapping circles. There are specific task, team and individual functions for the leader, but (s)he must concentrate on the small central overlap area of the three circles.
- The team process has inputs and outputs. Good teams have three main attributes: high task fulfilment, high team maintenance, and low self-orientation.
- In dealing with the task, the team and its individuals, a situational style of leadership must be adopted. This may follow the Tannenbaum and Schmidt, and Blanchard models through directing, coaching, and supporting to delegating.

Stages of team development

- When teams are put together, they pass through Tuckman's forming (awareness), storming (conflict), norming (co-operation), and performing (productivity) stages of development.
- Teams that go through these stages successfully become effective and display clear objectives and agreed goals, openness and confrontation, support and trust, co-operation and conflict, good decision-making, appropriate leadership, review of the team processes, sound relationships, and individual development opportunities.

Team roles and personality types

- Valuable work on team behaviour by Belbin has identified eight team roles: co-ordinator, shaper, plant, monitor/evaluator, implementor, resource investigator, teamworker, finisher.
- Eight people are not required for a team, but the roles, either as the main or secondary individual functions, should be present. Analysing existing groups and their performance or behaviour, using the team roles concept, can lead to improvement.

- The Belbin team roles have the merit of simplicity, but a more complete, understandable, helpful approach is provided by the Myers-Briggs Type Indicator (MBTI).
- The MBTI is based on individuals' preferences on four scales for giving and receiving 'energy' (extroversion-E or introversion-I), gathering information (sensing-S or intuition-N), making decisions (thinking-T or feeling-F) and handling the outer world (judging-J or perceiving-P).
- An individual's type is the combination and interaction of the four scales and can be assessed initially by completion of a simple questionnaire. There are sixteen types in all, which may be displayed for a team on a type table.
- The five As: for any of the teamwork models and theories, the individuals must become aware, need to accept, adopt and adapt, in order to act. A skilled facilitator is always necessary.

Implementing teamwork for quality improvement – the DRIVE model

- A structured approach to problem-solving is provided by the DRIVE model: define the problem, review the information, investigate the problem, verify the solution, and execute the change which is similar to the Z-shaped MBTI stepwise problem-solving process: S-N-T-F.
- After initial problems are solved, others should be tackled – successful solutions motivating new teams. In all cases teams should follow a disciplined approach to problem-solving, using proven techniques.
- Teamwork may mean a change in culture, which must be supported by management through its activities and behaviour.

Adding the teams to the TQM model

The third and final hard management necessity – the teams – are added to the tools and systems to complete the TQM model.

References

1 See references under *TQM through people and teamwork* heading in Bibliography, pp 000–0.
2 *Ibid.*
3 John Oakland, *Total Quality Management*, 2nd edition, Butterworth-Heinemann, 1993.

Discussion questions

1 The so-called process approach has certain implications for organisational structures. Discuss the main organisational issues influencing the involvement of people in process improvement.

2 Various TQM teamwork structures are advocated by many writers. Describe the role of the various 'quality teams' in the continuous improvement process. How can an organisation ensure that the outcome of teamwork is consistent with its mission?

3 Describe the various types of quality teams which should be part of a total quality programme. Explain the organisational requirements associated with these and give some indication of how the teams operate.

4 A large insurance company has decided that teamwork is to be the initial focus of its TQM programme. Describe the role of the Quality Council and Process Quality Teams in managing teamwork initiatives in quality improvement.

5 Explain the difference between Quality Improvement Teams and Quality Circles. What is their role in quality improvement activities?

6 Discuss some of the factors that may inhibit teamwork activities in a TQM programme.

7 Suggest an organisation for teamwork in a quality improvement programme and discuss how the important aspects must be managed, in order to achieve the best results from the use of teams. Describe briefly how the teams would proceed, including the tools they would use in their work.

8 Describe in full the various types of quality teams which are necessary in a total quality programme. Give some indication of how the teams operate at each level and using the 'DRIVE' model discuss the problem-solving approach that may be adopted.

9 Discuss the various models for teamwork within a total quality approach to business performance improvement. Explain through these models the role of the individual in TQM, and what work can be carried out in this area to help teams through the 'storming' stage of their development.

10 Teamwork is one of the key 'necessities' for TQM. John Adair's 'Action-Centred Leadership' model is useful to explain the areas which require attention for successful teamwork. Explain the model in detail showing your understanding of each of the areas of 'needs'. Pay particular attention to the needs of the individual, showing how a technique such as the Myers Briggs Type Indicator (MBTI) or Belbin's Team Roles may be useful here.

Part Five

TQM – The Implementation

All words, and no performance.
    Philip Massinger, 1583–1640 from 'The Unnatural Combat', ca 1619

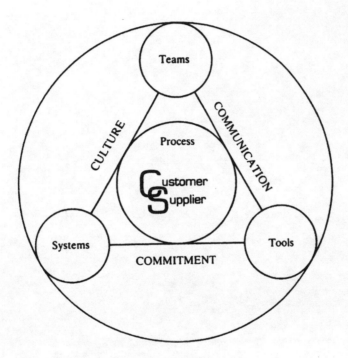

12

Communications and training for quality

12.1 Communicating the total quality strategy

People's attitudes and behaviour clearly can be influenced by communication; one has to look only at the media or advertising to understand this. The essence of changing attitudes to quality is to gain acceptance for the need to change, and for this to happen it is essential to provide relevant information, convey good practices, and generate interest, ideas and awareness through excellent communication processes. This is possibly the most neglected part of many organisations' operations, yet failure to communicate effectively creates unnecessary problems, resulting in confusion, loss of interest and eventually in declining quality through apparent lack of guidance and stimulus.

Total quality management will significantly change the way many organisations operate and 'do business'. This change will require direct and clear communication from the top management to all staff and employees, to explain the need to focus on processes. Everyone will need to know their roles in understanding processes and improving their performance.

Whether a strategy is developed by top management for the direction of the business/organisation as a whole, or specifically for the introduction of TQM, that is only half the battle. An early implementation step must be the clear widespread communication of the strategy.

An excellent way to accomplish this first step is to issue a total quality message that clearly states top management's commitment to TQM and outlines the role everyone must play. This can be in the form of a quality policy (see Chapter 2) or a specific statement about the organisation's intention to integrate TQM into the business operations. Such a statement might read:

> The Board of Directors (or appropriate title) believe that the successful implementation of Total Quality Management is critical to achieving and maintaining our business goals of leadership in quality, delivery and price competitiveness.
>
> We wish to convey to everyone our enthusiasm and personal commitment to the Total Quality approach, and how much we need your support in our mission of process improvement. We hope that you will become as convinced as we are that process improvement is critical for our survival and continued success.

We can become a Total Quality organisation only with your commitment and dedication to improving the processes in which you work. We will help you by putting in place a programme of education, training, and teamwork development, based on process improvement, to ensure that we move forward together to achieve our business goals.

The quality director or TQM co-ordinator should then assist the quality council to prepare a directive. This must be signed by all business unit, division, or process leaders, and distributed to everyone in the organisation. The directive should include the following:

- Need for improvement.
- Concept for total quality.
- Importance of understanding business processes.
- Approach that will be taken.
- Individual and process group responsibilities.
- Principles of process measurement.

The systems for disseminating the message should include all the conventional communication methods of seminars, departmental meetings, posters, newsletters, etc. First line supervision will need to review the directive with all the staff, and a set of questions and answers may be suitably pre-prepared in support.

Once people understand the strategy, the management must establish the infrastructure (see Chapter 10). The required level of individual commitment is likely to be achieved, however, only if everyone understands the aims and benefits of TQM, the role they must play, and how they can implement process improvements. For this understanding a constant flow of information is necessary, including:

1 When and how individuals will be involved.
2 What the process requires.
3 The successes and benefits achieved.

The most effective means of developing the personnel commitment required is to ensure people know what is going on. Otherwise they will feel left out and begin to believe that TQM is not for them, which will lead to resentment and undermining of the whole process. The first line of supervision again has an important part to play in ensuring key messages are communicated and in building teams by demonstrating everyone's participation and commitment.

Effective TQM communications, then, have two essential components:

(a) General information about the TQM process.
(b) Regular meetings between employees and managers/supervisors.

These equate respectively to the:

(a) 'Technical' aspects of the TQM framework or model.
(b) Human and organisational aspects of launching the whole process.

TQM will clearly have a profound effect on all tasks, activities, and processes throughout the organisation. It should change management style and integrate the process inputs of information, people, machines, and materials. One aspect of the communication process worthy of particular attention in this context is that between departments or functions. This is essential for establishing up-to-the-minute customer-oriented goals and building the 'house of quality' around the business processes.

The language used between departmental or functional groups will need attention in many organisations. Reducing the complexity and jargon in written and spoken communications will facilitate comprehension. When written business communications cannot be read or understood easily, they receive only cursory glances, rather than the detailed study they require. *Simplify and shorten* must be the guiding principles.

All levels of management should introduce and stress 'open' methods of communication, by maintaining open offices, being accessible to staff/employees, and taking part in day-to-day interactions and the detailed processes. This will lay the foundation for improved interactions *between* staff and employees, which is essential for information flow and process improvement. Opening these lines of communication may lead to confrontation with many barriers and much resistance. Training and the behaviour of supervisors/managements should be geared to helping people accept responsibility for their own behaviour, which often creates the barriers, and for breaking the barriers down by concentrating on the process rather than 'departmental' needs.

Resistance to change will always occur and is to be expected. Again first line management must be trained to help people deal with it. This requires an understanding of the dynamics of change and the support necessary – not an obsession with forcing people to change. Opening up lines of communication through a previously closed system, and publicising people's efforts to change and their results, will aid the process. Change can be – even should be – exciting if employees start to share their development, growth, suggestions, and questions. Management must encourage and participate in this by creating the most appropriate communication systems.

The key medium for motivating the employees and gaining their commitment to TQM is face-to-face communication and *visible* management commitment. Much is written and spoken about leadership, but it is mainly about communication. If people are good leaders, they are invariably good communicators. Leadership is a human interaction depending on the communications between the leaders and the followers. It calls for many skills that can be *learned* from education and training, but must be *acquired* through practice.

12.2 It's Monday – it must be training

It is the authors' belief that training is the single most important factor in actually improving quality, once there has been commitment to do so. For training to be effective, however, it must be planned in a systematic and objective manner. Quality training must be continuous to meet not only changes in technology but also changes in the environment in which an organisation operates, its structure, and perhaps most important of all the people who work there.

Training cycle of improvement

Quality training activities can be considered in the form of a cycle of improvement (Figure 12.1), the elements of which are the following.

Ensure training is part of the quality policy

Every organisation should define its policy in relation to quality (see Chapter 2). The policy should contain principles and goals to provide a framework within which training activities may be planned and operated. This policy should be communicated to all levels.

Allocate responsibilities for training

Quality training must be the responsibility of line management, but there are also important roles for the quality manager and his/her function.

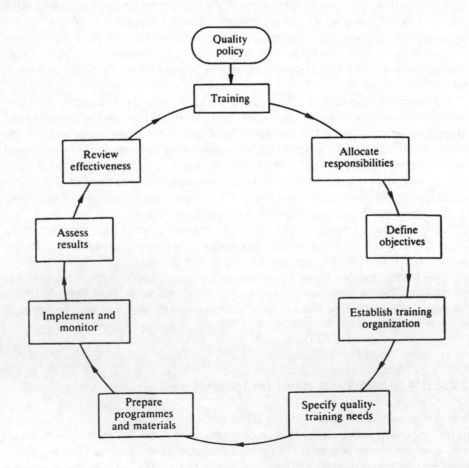

Figure 12.1   *The quality training circle*

Define training objectives

The following questions are useful first steps when identifying training objectives:

- How are the customer requirements transmitted through the organisation?
- Which areas need improved performance?
- What changes are planned for the future?
- What new procedures and provisions need to be drawn up?

When attempting to set training objectives three essential requirements must be met:

1 Senior management must ensure that objectives are clarified and priorities set.
2 Defined objectives must be realistic and attainable.
3 The main problems should be identified for all functional areas in the organisation. Large organisations may find it necessary to promote a phased plan to identify these problems.

Establish training organisation

The overall responsibility for seeing that quality training is properly organised must be assumed by one or more designated senior executives. All managers have a responsibility for ensuring that personnel reporting to them are properly trained and competent in their jobs. This responsibility should be written into every manager's job description. The question of whether line management requires specialised help should be answered when objectives have been identified. It is often necessary to use specialists, who may be internal or external to the organisation.

Specify quality training needs

The next step in the cycle is to assess and clarify specific quality training needs. The following questions need to be answered:

(a) Who needs to be trained?
(b) What competences are required?
(c) How long will training take?
(d) What are the expected benefits?
(e) Is the training need urgent?
(f) How many people are to be trained?
(g) Who will undertake the actual training?
(h) What resources are needed, eg money, people, equipment, accommodation, outside resources?

Prepare training programmes and materials

Quality management should participate in the creation of draft programmes, although line managers should retain the final responsibility for what is implemented, and they will often need to create the training programmes themselves.

Quality-training programmes should include:

- The training objectives expressed in terms of the desired behaviour.
- The actual training content.
- The methods to be adopted.
- Who is responsible for the various sections of the programme.

Implement and monitor training

The effective implementation of quality training programmes demands considerable commitment and adjustment by the trainers and trainees alike. Training is a progressive process, which must take into account the learning problems of the trainees.

Assess the results

In order to determine whether further training is required, line management should themselves review performance when training is completed. However good the quality training may be, if it is not valued and built upon by managers and supervisors, its effect can be severely reduced.

Review effectiveness of training

Senior management will require a system whereby decisions are taken at regular fixed intervals on:

- The quality policy.
- The quality training objectives.
- The training organisation.

Even if the quality policy remains constant, there is a continuing need to ensure that new quality training objectives are set either to promote work changes or to raise the standards already achieved.

The purpose of system audits and reviews is to assess the effectiveness of an organisation's quality effort. Clearly, adequate and refresher training in these methods is essential if such checks are to be realistic and effective. Audits and reviews can provide useful information for the identification of changing quality training needs.

The training organisation should similarly be reviewed in the light of the new objectives, and here again it is essential to aim at continuous improvement. Training must never be allowed to become static, and the effectiveness of the organisation's quality training programmes and methods must be assessed systematically.

12.3 A systematic approach to quality training

Training for quality should have, as its first objective, an appreciation of the personal responsibility for meeting the 'customer' requirements by everyone from the most senior executive to the newest and most junior employee. Responsibility for the training of employees in quality rests with management at all levels and, in particular, the person nominated for the co-ordination of the organisation's quality effort. Quality training will not be fully effective, however, unless responsibility for the quality policy rests clearly with the Chief Executive. One objective of this policy should be to develop a *climate* in which everyone is quality conscious and acts with the needs of the immediate customer in mind. Quality objectives should be stated in relation to the activities and the place of training in their achievement.

The main elements of effective and systematic quality training may be considered under four broad headings.

- Error/defect/problem prevention.
- Error/defect/problem reporting and analysis.
- Error/defect/problem investigation.
- Review.

The emphasis should obviously be on error, defect, or problem prevention, and hopefully what is said under the other headings maintains this objective.

Error/defect/problem prevention

The following contribute to effective and systematic training for prevention of problems in the organisation:

1 An issued quality policy.
2 A written quality system.
3 Job specifications that include quality requirements.
4 An effective quality council or committee including representatives of both management and employees.
5 Efficient housekeeping standards.
6 Preparation and display of flow diagrams and charts for all processes.

Error/defect/problem reporting and analysis

It will be necessary for management to arrange the necessary reporting procedures, and ensure that those concerned are adequately trained in these procedures. All errors, rejects, defects, defectives, problems, waste, etc, should be recorded and analysed in a way that is meaningful for each organisation, bearing in mind the corrective action programmes that should be initiated at appropriate times.

Error/defect/problem investigation

The investigation of errors, defects, and problems can provide valuable information that can be used in their prevention. Participating in investigations offers an opportunity for training. The following information is useful for the investigation:

(a) Nature of problem.
(b) Date, time and place.
(c) Product/service with problem.

(d) Description of problem.
(e) Causes and reasons behind causes.
(f) Action advised.
(g) Action taken to prevent recurrence.

Effective problem investigation requires appropriate follow-up and monitoring of recommendations.

Review of quality training

Review of the effectiveness of quality training programmes should be a continuous process. However, the measurement of effectiveness is a complex problem. One way of reviewing the content and assimilation of a training course or programme is to monitor behaviour during quality audits. This review can be taken a stage further by comparing employees' behaviour with the objectives of the quality training programme. Other measures of the training processes should be found to establish the benefits derived.

Training records

All organisations should establish and maintain procedures for the identification of training needs and the provision of the actual training itself. These procedures should be designed (and documented) to include all personnel. In many situations it is necessary to employ professionally qualified people to carry out specific tasks, eg accountants, lawyers, engineers, chemists, etc, but it must be recognised that all other employees, including managers, must have or receive from the company the appropriate education, training and/or experience to perform their jobs. This leads to the establishment of training records.

Once an organisation has identified the special skills required for each task, and developed suitable training programmes to provide competence for the tasks to be undertaken, it should prescribe how the competence is to be demonstrated. This can be by some form of examination, test or certification, which may be carried out in-house or by a recognised external body. In every case, records or personnel qualifications, training, and experience should be developed and maintained. National vocational qualifications (NVQs) have an important role to play here.

At the simplest level this may be a record of tasks and a date placed against each employee's name as he/she acquires the appropriate skill through training. Details of attendance on external short courses, in-house induction or training schemes complete such records. What must be clear and easily retrievable is the status of training and development of any single individual, related to the tasks that he/she is likely to encounter. For example, in a factory producing contact lenses that has developed a series of well defined tasks for each stage of the manufacturing process, it would be possible, by turning up the appropriate records, to decide whether a certain operator is competent to carry out a lathe-turning process. Clearly, as the complexity of jobs increases and managerial activity replaces direct manual skill, it becomes more difficult to make decisions on the basis of such records alone. Nevertheless, they should document the basic competency requirements and assist the selection procedure.

12.4 Starting where and for whom?

Training needs occur at four levels of an organisation:

- *Very senior management* (strategic decision-makers).
- *Middle management* (tactical decision-makers or implementors of policy).
- *First level supervision and quality team leaders* (on-the-spot decision-makers).
- *All other employees* (the doers).

Neglect of training in any of these areas will, at best, delay the implementation of TQM. The provision of training for each group will be considered in turn, but it is important to realise that an integrated training programme is required, one that includes follow-up activities and encourages exchange of ideas and experience, to allow each transformation process to achieve quality at the supplier/customer interface.

Very senior management

The Chief Executive and his team of strategic policy makers are of primary importance, and the role of training here is to provide awareness and instil commitment to quality. The importance of developing real commitment must be established; and often this can only be done by a free and frank exchange of views between trainers and trainees. This has implications for the choice of the trainers themselves, and the fresh-faced graduate, sent by the 'package consultancy' operator into the lion's den of a boardroom, will not make much impression with the theoretical approach that he or she is obliged to bring to bear. One of the authors recalls thumping many a boardroom table, and using all his experience and whatever presentation skills he could muster, to convince senior managers that without the TQM approach they would fail. It is a sobering fact that the pressure from competition and customers has a much greater record of success than enlightenment, although dragging a team of senior managers down to the shop floor to show them the results of poor management was successful on one occasion.

Executives responsible for marketing, sales, finance, design, operations, purchasing, personnel, distribution, etc must all be helped to understand quality. They must be shown how to define the quality policy and objectives, how to establish the appropriate organisation for quality, how to clarify authority, and generally how to create the atmosphere in which total quality will thrive. This is the only group of people in the organisation that can ensure that adequate resources are provided and they must be directed at:

1 Meeting customer requirements – internally and externally.
2 Setting standards to be achieved – zero failure.
3 Monitoring of quality performance – quality costs.
4 Introducing a good quality management system – prevention.
5 Implementing process control methods – SPC.
6 Spreading the idea of quality throughout the whole workforce – TQM.

Middle management

The basic objectives of management quality training should be to make managers conscious and anxious to secure the benefits of the total quality effort. One particular 'staff' manager will require special training – the quality manager, who will carry the responsibility for management of the quality system, including its design, operation, and review.

The middle managers should be provided with the technical skills required to design, implement, review, and change the parts of the quality system that will be under their direct operational control. It will be useful throughout the training programmes to ensure that the responsibilities for the various activities in each of the functional areas are clarified. The presence of a highly qualified and experienced quality manager must not allow abdication of these responsibilities, for the internal 'consultant' can easily create not-invented-here feelings by writing out procedures without adequate consultation of those charged with implementation.

Middle management must receive comprehensive training on the philosophy and concepts of teamwork, and the techniques and applications of statistical process control (SPC). Without the teams and tools, the quality system will lie dormant and lifeless. It will relapse into a paper generating system, fulfilling the needs of only those who thrive on bureaucracy.

First-level supervision

There is a layer of personnel in many organisations which plays a vital role in their inadequate performance – foremen and supervisors – the forgotten men and women of industry and commerce. Frequently promoted from the 'shop floor' (or recruited as graduates in a flush of conscience and wealth!), these people occupy one of the most crucial managerial roles, often with no idea of what they are supposed to be doing, without an identity, and without training. If this behaviour pattern is familiar and is continued, then TQM is doomed.

The first level of supervision is where the implementation of total quality is actually 'managed'. Supervisors' training should include an explanation of the principles of TQM, a convincing exposition on the commitment to quality of the senior management, and an explanation of what the quality policy means for them. The remainder of their training should then be devoted to explaining their role in the operation of the quality system, teamwork, SPC etc, and to gaining *their* commitment to the concepts and techniques of total quality.

It is often desirable to involve the middle managers in the training of first line supervision in order to:

- Ensure that the message they wish to convey through their tactical manoeuvres is not distorted.
- Indicate to the foreman level that the organisation's whole management structure is serious about quality, and intends that everyone is suitably trained and concerned about it too. One display of arrogance towards the training of supervisors and the workforce can destroy such careful planning, and will certainly undermine the educational effort.

All other employees

Awareness and commitment at the point of production or operation is just as vital as at the very senior level. If it is absent from the latter, the TQM programme will not begin; if it is absent from the shop floor, total quality will not be implemented. The training here must include the basics of quality, and particular care should be given to using easy reference points for the explanation of the terms and concepts. Most people can relate to quality and how it should be managed, if they can think about the applications in their own lives and at home. Quality is really such common sense that, with sensitivity and regard to various levels of intellect and experience, little resistance should be experienced.

All employees should receive detailed training in the quality procedures relevant to their own work. Obviously they must have appropriate technical or 'job' training, but they must also understand the requirements of their customers. This is frequently a difficult concept to introduce, particularly in the non-manufacturing areas, and time and follow-up assistance must be given if TQM is to take hold. It is always bad management to ask people to follow instructions without understanding why and where they fit into their own scheme of things.

12.5 Follow-up

For the successful implementation of TQM, training must be followed up during the early stages. Follow-up can take many forms, but the managers must provide the lead through the design of improvement projects and 'surgery' workshops.

In introducing statistical methods of process control, for example, the most satisfactory strategy is to start small and build up a bank of knowledge and experience. Sometimes it is necessary to introduce SPC techniques alongside existing methods of control (if they exist) thus allowing comparisons to be made between the new and old methods. When confidence has been established from these comparisons, the SPC methods will almost take over the control of the processes themselves. Improvements in one or two areas of the organisation's operations, by means of this approach will quickly establish the techniques as reliable methods of controlling quality.

The authors and their colleagues have found that a successful formula is the in-company training course plus follow-up workshops. Usually a 20-hour seminar on TQM is followed within a few weeks by an 8–10 hour workshop at which participants on the initial training course present the results of their efforts to improve processes, and use the various methods. The presentations and specific implementation problems may be discussed. A series of such workshops will add continually to the follow-up, and can be used to initiate quality improvement teams. Wider company presence and activities should be encouraged by the follow-up activities.

Chapter highlights

Communicating the total quality strategy

- People's attitudes and behaviour can be influenced by communication, and the essence of changing attitudes is to gain acceptance through excellent communication processes.

- The strategy and changes to be brought about through TQM must be clearly and directly communicated from top management to all staff/employees. The first step is to issue a 'total quality message'. This should be followed by a signed TQM directive.
- People must know when and how they will be brought into the TQM process, what the process is, and the successes and benefits achieved. First-line supervision has an important role in communicating the key messages and overcoming resistance to change.
- The complexity and jargon in the language used between functional groups must be reduced in many organisations. Simplify and shorten are the guiding principles.
- 'Open' methods of communication and participation must be used at all levels. Barriers may need to be broken down by concentrating on process rather than 'departmental' issues.
- Good leadership is mostly about good communications, the skills of which can be learned through training but must be acquired through practice.

It's Monday – it must be training

- Training is the single most important factor in improving quality, once commitment is present. Quality training must be objectively, systematically, and continuously performed.
- All training should occur in an improvement cycle of ensuring training is part of quality policy, allocating responsibilities, defining objectives, establishing training organisations, specifying needs, preparing programmes and materials, implementing and monitoring, assessing results, and reviewing effectiveness.

A systematic approach to quality training

- Responsibility for quality training of employees rests with management at all levels. The main elements should include error/defect/problem prevention, reporting and analysis, investigation, and review.
- Training procedures and records should be established. These should show how job competence is demonstrated.

Starting where and for whom?

- Needs for integrating quality training occur at four levels of the organisation: very senior management, middle management, first level supervision and quality team leaders, and all other employees.

Follow-up

- All quality training should be followed up with improvement projects and 'surgery' workshops.

13

Implementation of TQM and the management of change

13.1 TQM and the management of change

The authors recall the managing director of a large transportation company who decided that a major change was required in the way the company operated if serious competitive challenges were to be met. The Board of Directors went away for a weekend and developed a new vision for the company and its 'culture'. A personnel director was recruited and given the task of managing the change in the people and their 'attitudes'. After several 'programmes' aimed at achieving the required change, including a new structure for the organisation, a staff appraisal system linked to pay, training programmes to change attitudes, and questionnaire surveys, very little change in actual organisational behaviour had occurred.

Clearly something had gone wrong somewhere. But what, who, where? Everything was wrong, including who needed changing, who should lead the changes and, in particular, how the changes should be brought about. This type of problem is very common in organisations desiring to change the way they operate to deal with increased competition, a changing market place, and different business rules. In this situation many companies recognise the need to move away from an autocratic management style, with formal rules and hierarchical procedures, and narrow work demarcations. Some have tried to create teams, to delegate (perhaps for the first time), and to improve communications.

Some of the senior managers in such organisations recognise the need for change to deal with the new realities of competitiveness, but they lack an understanding of how the change should be implemented. They often believe that changing the formal organisational structure, having vision or mission statements, 'culture change' programmes, training courses, and new payment systems will, by themselves, make the transformations.

In much research work carried out in the United Kingdom, the United States and Australia, it has been shown that there is almost an inverse relationship between successful change and having formal organisation-wide change. This is particularly true if one functional group, such as personnel, 'owns' the programme.

In several large organisations in which total quality has been used successfully to effect change, the senior management did not focus on formal structures and systems, but set up *process-management* teams to solve real business or organisation problems.

The key to success in this area is to align the employees of the business, their roles and responsibilities with the organisation and its *processes*. This is the core of process mapping or alignment. When an organisation focuses on its key processes, that is the activities and tasks themselves, rather than on abstract issues such as 'culture' and 'participation', then the change process can begin in earnest.

An approach to change, based on process alignment, and starting with the mission statement, analysing the critical success factors, *and* moving on to the key or critical processes, is the most effective way to engage the staff in an enduring change process. Many change programmes do not work because they begin trying to change the knowledge, attitudes and beliefs of individuals. The theory is that changes in these areas will lead to changes in behaviour throughout the organisation. It relies on a form of religion spreading through the people in the business.

What is required, however, is virtually the opposite process, based on the recognition that people's behaviour is determined largely by the roles they have to take up. If we create for them new responsibilities, team roles, and a process driven environment, a new situation will develop, one that will force their attention and work on the processes. This will change the culture. *Teamwork* is an especially important part of the TQM model in terms of bringing about change. If changes are to be made in quality, costs, market, product or service development, close co-ordination among the marketing, design, production/operations and distribution groups is essential. This can be brought about effectively only by multifunctional teams working on the processes and understanding their interrelationships. *Commitment* is a key element of support for the high levels of co-operation, initiative, and effort that will be required to understand and work on the labyrinth of processes existing in most organisations. In addition to the knowledge of the business as a whole, which will be brought about by an understanding of the mission → CSF → process breakdown links, certain *tools, techniques*, and *interpersonal skills* will be required for good *communication* around the processes. These are essential if people are to identify and solve problems as teams.

If any of these elements are missing the total quality underpinned change process will collapse. The difficulties experienced by many organisations' formal change processes is that they tackle only one or two of these necessities. Many organisations trying to create a new philosophy based on teamwork fail to recognise that the employees do not know which teams to form or how they should function as teams. Recognition that effective teams need to be formed round their process, which they begin to understand together – perhaps for the first time – and further recognition that they then need to be helped as individuals through the forming-storming-norming-performing sequence, will generate the interpersonal skills and attitude changes necessary to make the new 'structure' work.

13.2 Integrating TQM into the strategy of the business

Organisations will avoid the problems of 'change programmes' by concentrating on 'process alignment' – recognising that people's roles and responsibilities must be related to the processes in which they work. Senior managers may begin the task of process alignment by a series of seven distinct but clearly overlapping steps. This recommended path develops a self-reinforcing cycle of *commitment, communication,* and *culture* change. The order of the steps is important because some of the activities will be inappropriate if started too early. In the introduction of total quality for managing change, timing can be critical.

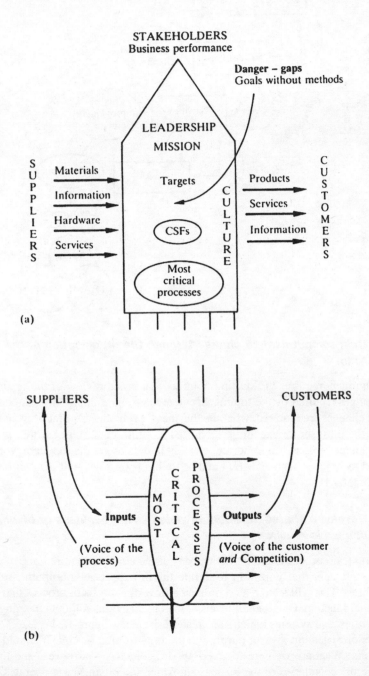

Figure 13.1   *(and opposite) From mission to process breakdown*

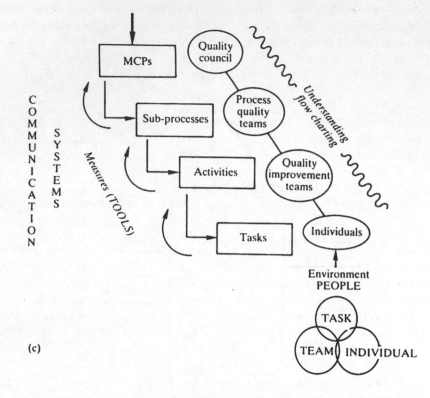

(c)

Step 1   Gain commitment to change through the organisation of the top team

Process alignment requires the starting point to be a broad review of the organisation and the changes required by the top management team. By gaining this shared diagnosis of what changes are required, what the 'business' problems are, and/or what must be improved, the most senior executive mobilises the initial commitment that is vital to begin the change process. An important element here is to get the top team working as a team, and techniques such as MBTI and/or Belbin team roles will play an important part (see Chapter 11).

Step 2   Develop a shared 'mission' or vision of the business or of what change is required

Once the top team is committed to the analysis of the changes required, it can develop a mission statement that will help to define the new process alignment, roles and responsibilities. This will lead to a co-ordinated flow of analysis of process that crosses the traditional functional areas at all levels of the organisation, without changing formal structures, titles, and systems which can create resistance (Figure 13.1).

The mission statement gives a purpose to the organisation or unit. It should answer the questions 'What are we here for?' or 'What is our basic purpose?' and therefore must define the boundaries of the business in which the organisation operates (Figure 13.1a). This will help to focus on the 'distinctive competence' of the organisation, and to orient everyone in the direction of what has to be done. The mission must be documented, agreed by the top management team, sufficiently explicit to enable its eventual accomplishment to be verified, and ideally be no more than four sentences. The statement must be understandable, communicable, believable, and usable.

Some questions that may be asked of a mission statement are:

- Does it contain the need that is to be fulfilled?
- Is the need worthwhile in terms of admiration and identification, both internally and externally?
- Does it take a long term view, leading to, for example, commitment to new product or service development, or training of personnel?
- Does it take into account all the 'stakeholders'?
- Will the purpose remain constant despite changes in top management?

It is important to establish in some organisations whether or not the mission is survival. This does not preclude a longer term mission, but the short term survival mission must be expressed, if it is relevant. The management team can then decide whether it wishes to continue long term strategic thinking. If survival is a real issue, the authors would advise against concentrating on the long term planning initially.

There must be open and spontaneous discussion during generation of the mission, but there must in the end be convergence on one statement. If the mission statement is wrong, everything that follows will be wrong too, so a clear understanding is vital.

Step 3 Define the measurable objectives, which must be agreed by the team, as being the quantifiable indicators of success in terms of the mission

The mission provides the vision and guiding light and sets down the core values, but it must be supported by measurable objectives that are tightly and unarguably linked to it. These will help to translate the directional and sometimes 'loose' statements of the mission into clear targets, and in turn to simplify management's thinking. They can later be used as evidence of success for the team, in every direction, internally and externally.

Step 4 Develop the mission into its critical success factors (CSFs) to coerce and move it forward

The development of the mission is clearly not enough to ensure its implementation. This is the 'danger gap' into which many companies fall, because they do not foster the skills needed to translate the mission through its CSFs into the critical processes. Hence they have 'goals without methods', and TQM is not integrated properly into the business. At this stage of the process strong leadership from the top is crucial. Commitment to the change, whatever it may be, is always imbalanced; some senior managers may be antagonistic, some neutral, others enthusiastic or worried about the proposed changes.

Once the top managers begin to list the CSFs, they will gain some understanding of what the mission or the change requires. The first step in going from mission to CSFs is to brainstorm all the possible impacts on the mission. In this way thirty to fifty items, ranging from politics to costs, from national cultures to regional market peculiarities, may be derived.

The CSFs may now be defined – what the organisation must accomplish to achieve the mission, by examination and categorisation of the impacts. There should be no more than eight CSFs, and no more than four if the mission is survival. They are the minimum key factors or subgoals that the organisation *must have* or *need*, and which together will achieve the mission. They are not the how, and are not directly manageable – they may be in some cases statements of hope or fear – but they provide direction and the success criteria. In CSF determination a management team should follow the rule that each CSF is *necessary*, and that together they are *sufficient* for the mission to be achieved.

Some examples of CSFs may clarify understanding.

* We must have right-first-time suppliers.
* We must have motivated, skilled people.
* We need new products that satisfy market needs.
* We need new business opportunities.
* We must have best-in-the-field product quality.

The list of CSFs should be an agreed balance of strategic and tactical issues, each of which deals with a 'pure' factor, the use of *and* being forbidden. It will be important to know when the CSFs have been achieved through Key Performance Indicators (KPIs), but the more important next step is to use the CSFs to enable the identification of the *processes*.

Step 5   Break down the critical success factors into the key or critical process and gain process ownership

This is the point at which the top management team have to consider how to institutionalise the mission or the change in the form of processes that will continue to be in place, after any changes have been effected (Figure 13.1b).

The key, critical, or business processes describe what actually is or needs to be done so that the organisation meets its CSFs. As with the CSFs and the mission, each process *necessary* for a given CSF must be identified, and together the processes listed must be *sufficient* for the CSFs to be accomplished. To ensure that *processes* are listed, they should be in the form of verb plus object, such as 'research the market', 'recruit competent staff', or 'measure supplier performance'.

Each business process should have an owner who is a member of the management team that agree the CSFs. The business processes identified frequently run across 'departments' or functions, yet they must be measurable.

The questions will now come thick and fast. Is the process currently carried out? By whom? When? How frequently? With what performance and how well compared with competitors? The answers to these will force process ownership into the business. The process owner should form a process quality team to take the next steps in quality improvement. Some form of prioritisation, by means of process 'quality' measures, is necessary at this stage to enable effort to be focused on the key areas for improvement. This may be carried out by a form of matrix analysis[1] or some other means. The outcome should be a set of 'most critical processes' (MCPs), which receive priority attention for improvement.

The first stage in understanding the critical processes is to produce a set of processes of a common order of magnitude. Some processes identified by the quality council may break into two or three critical processes; others may be already at the appropriate level. This method will ensure that the change becomes entrenched, the critical processes are identified and that the right people are in place to own or take responsibility for them; and it will be the start of getting the process-team organisation up and running.

Step 6   *Break down the critical processes into sub-processes, activities and tasks and form improvement teams around these*

Once an organisation has defined and mapped out the critical processes, people need to develop the skills to understand how the new process structure will be analysed and made to work. The very existence of new process quality teams (PQTs) with new goals and responsibilities will force the organisation into a learning phase. The changes should foster new attitudes and behaviours.

An illustration of the breakdown from mission through CSFs and critical processes to individual tasks may assist in understanding the process required:

> | Mission |

Two of the statements in a well known quality management consultancy's mission statement are: 'Gain and maintain a position as Australia's foremost management consultancy in the development of organisations through the management of change' and 'provide the consultancy, training and facilitation necessary to assist with making the continuous improvement of quality an integral part of our customers' business strategy'.

↓

> | Critical success factor |

One of the CSFs that clearly relates to this is 'We need a high level of awareness of our company in the market place'.

↓

> | Critical process |

One of the critical processes that clearly must be done particularly well to achieve this CSF is to 'Promote, advertise, and communicate the company's business capability'.

↓

> | Sub-process |

One of the sub-processes resulting from a breakdown of this critical process is 'Prepare the company's information pack'.

↓

One of the activities contributing to this sub-process is 'Prepare *one* of the subject booklets, ie TQM, SPC or quality systems'.

↓

Task

One of the tasks that contributes to this is 'Write the detailed leaflet for any particular seminar', eg: 'One-day or three-day seminars on TQM or SPC, or quality system advisory project'.

Individuals, tasks, and teams

Having broken down the processes into sub-processes, activities, and tasks in this way, we can now link them with the Adair model of action-centred leadership and teamwork.

The *tasks* are performed, at least initially, by individuals. For example, some*body* has to sit down and draft out the first version of a seminar leaflet. There has to be an understanding by the individual of the task and its position in the hierarchy of processes. Once the initial task has been performed, the results must be checked against the activity of co-ordinating the promotional booklet – say for TQM. This clearly brings in the team, and there must be interfaces between the needs of the *tasks,* the *individuals* who performed them and the *team* concerned with the *activities*.

Using the hierarchy of processes, it is possible to link this with the hierarchy of quality teams. Hence:

Quality council – mission – CSFs – critical processes.
Process quality teams – critical processes.
Quality improvement (or functional) teams (QITs) – sub-processes.
QITs – activities.
QITs and quality circles/Kaizen teams/individuals – tasks.

Performance measurement and metrics

Once the processes have been analysed in this way, it should be possible to develop metrics for measuring the performance of the processes, sub-processes, activities, and tasks. These must be meaningful in terms of the inputs and outputs of the processes, and in terms of the customers and of suppliers to the processes (Figure 13.1c).

At first thought, this form of measurement can seem difficult for processes such as preparing a sales brochure or writing leaflets advertising seminars, but, if we think carefully about the *customers* for the leaflet-writing tasks, these will include the *internal* ones, ie the consultants, and we can ask whether the output meets their requirements. Does it really say what the seminar is about, what its objectives are and what the programme will be? Clearly, one of the 'measures' of the seminar leaflet-writing task could be the number of typing errors in it, but is this a *key* measure of the performance of the process? Only in the context of office management is this an important measure. Elsewhere it is not.

The same goes for the *activity* of preparing the subject booklet. Does it tell the 'customer' what TQM or SPC is and how the consultancy can help? For the *sub-process* of preparing the company brochure, does it inform people about the company and does it bring in enquiries from which customers can be developed? Clearly, some of these measures require *external market research*, and some of them *internal research*. The main point is that metrics must be developed and used to reflect the *true performance* of the processes, sub-processes, activities, and tasks. These must involve good contact with external and internal customers of the processes. The metrics may be quoted as *ratios*, eg number of customers derived per number of brochures mailed out. Good data-collection, record-keeping, and analysis are clearly required.

It is hoped that this illustration will help the reader to:

• Understand the breakdown of processes into sub-processes, activities, and tasks.
• Understand the links between the process breakdowns and the task, individual and team concepts.
• Link the hierarchy of processes with the hierarchy of quality teams.
• Begin to assemble a cascade of flowcharts representing the process breakdowns, which can form the basis of the quality system and communicate what is going on throughout the business.
• Understand the way in which metrics must be developed to measure the true performance of the process, and their links with the customers, suppliers, inputs and outputs of the processes.

This whole concept/structure is represented in Figure 13.1c. The changed patterns of co-ordination, driven by the process maps, should increase collaboration and information sharing.

Clearly the senior and middle managers must provide the right support. Once employees, at all levels, identify what kinds of new skill are needed, they will ask for the formal training programmes in order to develop those skills further. This is a key area, because the teamwork around the processes will ask more of employees, so they will need increasing support from their managers.

This has been called 'just-in-time' training, which describes very well the nature of the process required. Such training is quite different from the blanket or carpet-bombing training associated with many unsuccessful change programmes, which targets competencies or skills but does not change the organisation's patterns of collaboration and co-ordination.

Step 7   Monitor and adjust the process alignment in response to difficulties in the change process

Change must create something that did not exist before, namely a 'learning organisation' capable of adapting to a changing competitive environment. One must also learn how to monitor and modify the new behaviour to maintain the change-sensitive environment.

Some people will, of course, find great difficulty in accepting the changes, and perhaps will be incapable of doing so, in spite of all the direction, support, and peer pressure brought about by the process alignment. There will come a time to replace

```
C        Initial key steps following one day seminar for top
         management
O        • Formation of quality council (Top-management
           team)
M        • TQM 'Attitude' survey
         – Profile of organization
M    C   – Quality costs
         – Strengths/weaknesses
I    O   • Two-day strategic planning workshop (quality
           council)
T    M   – Charter
         – Mission statement
M    M   – Quality policy
         – Critical success factors
E    U   – Critical processes
         – Implementation action plan
N    N   • Formation of process quality teams and/or site
           steering committees
T    I   • Teamwork seminar for quality council (may
           precede strategic planning workshop)
     C   • Identify team facilitators
   A C   • Run specific training and team-forming workshops
   T U   • Company-wide awareness training on
           customer/supplier interfaces
   I L   • Implementation/improvement projects for quality
           policy deployment
   O T   – Quality costing
         – Customer/supplier framework
   N U   – DPA
         – Systems
     R   – Techniques
     E   • Feedback/follow-up workshops throughout
           implementation
```

Figure 13.2   *TQM implementation*

those managers and staff who cannot function in the new organisation, after they have had a good opportunity to make the changes. These decisions are of course never easy, especially where valuable technical skills are owned by the people who have difficulty working in the new participatory, process-driven organisation.

When people begin to understand what kind of manager and employee/staff the new organisation needs, and this often develops slowly and from experience of seeing individuals succeed and fail, they should begin to accept the need to replace or move people to other parts of the organisation.

13.3 Summarising the steps

If a top-management team has attended at least a 1-day workshop on TQM, the initial key steps may be summarised in Figure 13.2 together with their links to commitment, communications and culture change, through project work and correct follow-up. This must all be done within the continuous improvement cycle to avoid the 'danger gaps' shown in Figure 13.3

13.4 Continuous improvement

Never-ending or continuous improvement is probably the most powerful concept to guide management. It is a term not well understood in many organisations, although that must begin to change if those organisations are to survive. To maintain a wave of interest in quality, it is necessary to develop generations of managers who not only understand but are dedicated to the pursuit of never-ending improvement in meeting external and internal customer needs.

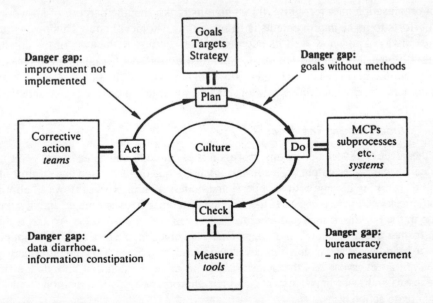

Figure 13.3 *TQM implementation – all done with the Deming continuous improvement cycle*

The concept requires a systematic approach to quality management that has the following components:

- *Planning* the processes and their inputs.
- *Providing* the inputs.
- *Operating* the processes.
- *Evaluating* the outputs.
- *Examining* the performance of the processes.
- *Modifying* the processes and their inputs.

This system must be firmly tied to a continuous assessment of customer needs, and depends on a flow of ideas on how to make improvements, reduce variation, and generate greater customer satisfaction. It also requires a high level of commitment, and a sense of personal responsibility in those operating the processes.

The never-ending improvement cycle ensures that the organisation learns from results, standardises what it does well in a documented quality management system, and improves operations and outputs from what it learns. But the emphasis must be that this is done in a planned, systematic, and conscientious way to create a climate – a way of life – that permeates the whole organisation.

There are three basic principles of never-ending improvement:

- Focusing on the *customer*.
- Understanding the *process*.
- All *employees* committed to quality.

1   Focusing on the customer

An organisation must recognise, throughout its ranks, that the purpose of all work and all efforts to make improvements is to serve the customers better. This means that it must always know how well its outputs are performing, in the eyes of the customer, through measurement and feedback. The most important customers are the external ones, but the quality chains can break down at any point in the flows of work. Internal customers therefore must also be well served if the external ones are to be satisfied.

2   Understanding the process

In the successful operation of any process it is essential to understand what determines its performance and outputs. This means intense focus on the design and control of the inputs, working closely with suppliers, and understanding process flows to eliminate bottlenecks and reduce waste. If there is one difference between management/supervision in the Far East (Japan and other leading Asian countries) and the West, it is that in the former management is closer to, and more involved in, the processes. It is not possible to stand aside and manage in never-ending improvement. TQM in an organisation means that everyone has the determination to use their detailed knowledge of the processes and make improvements, and use appropriate statistical methods to analyse and create action plans.

3   *All employees committed to quality*

Everyone in the organisation, from top to bottom, from offices to technical service, from headquarters to local sites, must play their part. People are the source of ideas and innovation, and their expertise, experience, knowledge, and co-operation have to be harnessed to get those ideas implemented.

When people are treated like machines, work becomes uninteresting and unsatisfying. Under such conditions it is not possible to expect quality services and reliable products. The rates of absenteeism and of staff turnover are measures that can be used in determining the strengths and weaknesses, or management style and people's morale, in any company.

The first step is to convince everyone of their own role in total quality. Employers and managers must of course take the lead, and the most senior executive has a personal responsibility for quality. The degree of management's enthusiasm and drive will determine the ease with which the whole workforce is motivated.

Most of the work in any organisation is done away from the immediate view of management and supervision, and often with individual discretion. If the co-operation of some or all of the people is absent, there is no way that managers will be able to cope with the chaos that will result. This principle is extremely important at the points where the processes 'touch' the outside customer. Every phase of these operations must be subject to continuous improvement, and for that everyone's co-operation is required.

Never-ending improvement is the process by which greater customer satisfaction is achieved. Its adoption recognises that quality is a moving target, but its operation actually results in quality.

13.5 A model for total quality management

The concept of total quality management is basically very simple. Each part of an organisation has customers, whether within or without, and the need to identify what the customer requirements are, and then set about meeting them, forms the core of a total quality approach. This requires the three hard management necessities: a good quality management system, tools such as statistical process control (SPC), and teamwork. These are complementary in many ways, and they share the same requirement for an uncompromising commitment to quality. This must start with the most senior management and flow down through the organisation. Having said that, teamwork, SPC, or the quality system, or all three, may be used as a spearhead to drive TQM through an organisation. The attention to many aspects of a company's operations – from purchasing through to distribution, from data recording to control chart plotting – which are required for the successful introduction of a good quality system, or the implementation of SPC, will have a 'Hawthorne effect', concentrating everyone's attention on the customer-supplier interface, both inside and outside the organisation.

Total quality management calls for consideration of processes in all the major areas: marketing, design, procurement, operations, distribution, etc. Clearly, these each require considerable expansion and thought, but if attention is given to all areas, using the concepts of TQM, then very little will be left to chance. Much of industry and commerce would benefit from the improvements in quality brought about by the

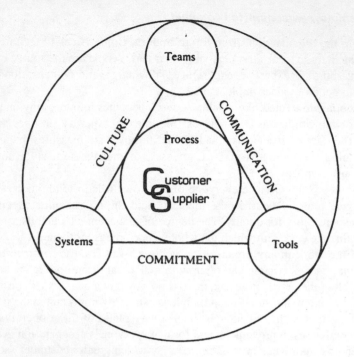

Figure 13.4   *Total quality management model*

approach represented in Figures 13.1 and 13.4. This approach will ensure the implementation of the management commitment represented in the quality policy, and
provide the environment and information base on which teamwork thrives.

Chapter highlights

TQM and the management of change

*   Senior managers in some organisations recognise the need for change to deal with
    increasing competitiveness, but lack an understanding of how to implement the
    changes.
*   Successful change is effected not by focusing on formal structures and systems, but
    by aligning process management teams. This starts with writing the mission state-
    ment, analysis of the critical success factors (CSFs) and understanding the critical
    or key processes.

Integrating TQM into the strategy of the business/the steps

*   Senior management may begin the task of process alignment through seven steps to
    a self-reinforcing cycle of commitment, communication, and culture change.
*   The first three steps are gain commitment to change, develop a shared mission or
    vision of the business or desired change, and define the measurable objectives.

- The remaining four steps comprise developing the mission into its CSFs; understanding the key or critical processes and gaining ownership; breaking down the critical processes into sub-processes, activities and tasks; and monitoring and adjusting the process alignment in response to difficulties in the change process.

Continuous improvement

- Managers must understand and pursue never-ending improvement. This should cover planning and operating processes, providing inputs, evaluating outputs, examining performance, and modifying processes and their inputs.
- There are three basic principles of continuous improvement: focusing on the customer, understanding the process, and seeing that all employees are committed to quality.

A model for TQM

- In the model for TQM the customer–supplier chains form the core, which is surrounded by the hard management necessities of a good quality system, tools, and teamwork.

Reference

1   See, for example, Hardaker, M and Ward, B K, 'Getting Things done – how to make a team work', *Harvard Business Review*, Nov/Dec 1987, pp 112–119.

Discussion questions

1 You have just been appointed as the Production Manager of a small chemical company. You are shocked at the apparent disregard for procedures which have been laid down. This is particularly noticeable among the younger/newer members of the work force.

   Briefly outline your responsibility in the area of quality and describe how you could proceed to improve the situation.

2 You have recently been appointed as Transport Manager of the haulage division of an expanding company and have been alarmed to find that maintenance costs seem to be higher than you would have expected in an efficient organisation.

   Outline some of the measures that you would take to bring the situation under control.

3 TQM has been referred to as 'a rain dance to make people feel good without impacting on bottom line results'. It was also described as 'flawed logic that confuses ends with means, processes with outcomes'. The arguments on whether to focus on budget control through financial management or quality improvement through process management clearly will continue in the future.

   Discuss the problems associated with taking a financial management approach which has been the traditional method used in the West.

4 (a) Discuss what is meant by taking a process management approach. What are the key advantages of focusing on process improvement?
   (b) Discuss how TQM can impact on bottom line results.

5 What are the critical elements of integrating total quality management or business improvement into the strategy of an organisation? Illustrate your approach with reference to an organisation with which you are familiar, or which you have heard about and studied.

6 You are the new Quality Director of part of ICI. Some members of the top management team have had some brief exposure to TQM, and you have been appointed to lay down plans for its implementation.

   Set down plans for the process which you would initiate to achieve this. Your plans should include reference to any training needs, outside help and additional internal appointments required, with timescales.

7 You are the new Quality Director of ONE of the following:
   Singapore Airlines, Royal Melbourne Hospital or Monash University

   The members of your top management team have had some brief exposure to TQM and you have been appointed to lay down plans for its implementation.

   Choose any of the above organisations and set down plans for the process which you would initiate to achieve this. Your plans should be as fully developed as possible in the time allowed and include reference to any training needs, outside help and additional internal appointments required, with a realistic timescale.

TQM
Case Studies

Introduction to case studies

Reading, using, analysing the cases

The cases in this book provide a description of what occurred in 11 different organisations from the Pacific Rim region, regarding various aspects of their quality improvement efforts. They may each be used as a learning vehicle as well as providing information and description which demonstrate the application of the concepts and techniques of TQM.

The objective of writing the cases has been to offer a resource through which the student of TQM (including many practising managers) understands how TQM companies operate. We hope that the cases will provide a useful and distinct contribution to TQM education and training.

The case material is suitable for practising managers, students on undergraduate and postgraduate courses, and all teachers of the various aspects of business management and TQM.

The cases have been written so that they may be used in three ways:

1 As orthodox cases for student preparation and discussion.
2 As illustrations, which teachers may also use as support for their other methods of training and education.
3 As supporting/background reading on TQM.

If used in the orthodox way, it is recommended that firstly the case is read to gain an understanding of the issues and to raise questions which may lead to a collective and more complete understanding of the company, TQM and the issues in the particular case. Secondly, case discussion or presentations in groups will give practise in putting forward thoughts and ideas persuasively.

The greater the effort put into case study, preparation, analysis and discussion in groups, the greater will be the individual benefit. There are, of course, no correct and tidy cases in any subject area. What the directors and managers of an organisation actually did is not necessarily the best way forward. One object of the cases is to make the reader think about the situation, the problems and the progress made, and what improvements or developments are possible.

The writing of each case emphasises particular problems or issues which were apparent for the organisation. This may have obscured other more important ones. The diagnostic skill of the student will allow the symptoms to be separated from the disease. Imagination, innovation and intuition should be as much a part of the study of a case as observation and analysis of the facts and any data available.

TQM cases, by their nature, will be very complicated and, to render the cases in this book useful for study, some of the complexity has been omitted. This simplification is accompanied by the danger of making the implementation seem clear-cut and obvious. Believe us, that is never the case with TQM!

TQM case analysis

The main objective of each description is to enable the reader to understand the situation and its implications, and to learn from the particular experiences. The cases are not, in the main, offering specific problems to be solved. In using the cases, the reader/student should try to:

- *Recognise or imagine* the situation in the organisation.
- *Understand* the context and objectives of the process(es) described.
- *Analyse* the different parts of the case (including any data) and their interrelationships.
- *Determine* the overall structure of the situation/problem(s)/case.
- *Consider* the different options facing the organisation.
- *Evaluate* the options and the course of action chosen, using any results stated.
- *Consider any recommendations* which should be made to the organisation for further work, action, or implementation.

   The set of cases has been chosen to provide a good coverage across different types of industry and organisation, including those in the service, manufacturing and public sectors.

   The value of illustrative cases in an area such as TQM is that they inject reality into the conceptual frameworks developed by authors in the subject. The cases are all based on real situations and are designed to illustrate certain aspects of managing change in organisations, rather than to reflect good or poor management practice. The cases may be used for analysis, discussion, evaluation, and even decision-making within groups without risk to the individuals, groups, or organisation(s) involved. In this way students of TQM may become 'involved' in many different organisations and their approaches to TQM implementation, over a short period and in an efficient and effective way.

   The organisations described here have faced translating TQM theory into practice, and the description of their experiences should provide an opportunity for the reader of TQM literature to test his/her preconceptions and understanding of this topic. All the cases describe real TQM processes in real organisations and we are grateful to the people involved for their contribution to this book.

Further reading

Easton, G, *Learning from Case Studies* (2nd edn), Prentice-Hall, UK, 1992.

C1

The Quest for Quality at Safeway Australia

The company: Safeway Australia

Safeway, a division of Woolworths, is the largest supermarket chain in Victoria, Australia. Its head office and the largest warehouse is located in the Melbourne suburb of Mulgrave. Safeway has a total of 132 stores throughout the state, and employs over 17,000 employees. More than 10,000 items are sold in their supermarkets, ranging from groceries, deli products, fresh fruits and vegetables, and household goods. Safeway competes on the basis of service, atmosphere, low prices and more particularly on availability of fresh foods.

A total of eight employees, from general manager to checkout operator, were interviewed at the head office and warehouse located in Mulgrave, and at one of the stores located in Eltham. Some individuals were interviewed more than once. The information in this case study is based on those interviews. The principal contact was the quality manager, Margaret Cemm.

Total quality management at Safeway Australia

Safeway's involvement in Total Quality Management (TQM) began in the late 1980s. It was first introduced to TQM through the success of Walmart, a chain of 150 departmental stores in the United States. A consultant who had worked with the Walmart chain met with the managing director, Mr Harry Watts, and the senior management team of Big W, a division of Woolworths, in Sydney. At the meeting they were introduced to TQM, and the factors which contributed to Walmart's success. The managing director in turn conveyed his experiences to the general manager of Safeway, Mr Trevor Herd.

In 1989, after one of Safeway's country stores had reported losses for 12 consecutive years, and with no substantial action taken to rectify the situation, the general manager resolved to investigate its circumstances personally. What he found was a store with high levels of wastage, poor morale and rife pilfering of merchandise. He called a meeting in which all store employees were invited to attend, and presented to them for the first time the complete financial position of the store. When someone queried why the store had not been closed after so many unprofitable years, the general manager replied that, apart from any contractual obligations to keep the store open, he would like to know where everyone would have turned to if it were closed. It was this commitment which prompted staff to establish improvement teams within the store. Six months later the store reported a $60,000 profit, followed by a $120,000 profit after 12 months.

The success of this initiative prompted the general manager to appoint Margaret Cemm as Quality Manager for Safeway. A local consultant was also engaged to introduce the philosophy of TQM to senior managers. This was achieved through a one-day 'vision setting workshop' for senior managers.

Following the senior managers' workshop, Safeway sent all of their store managers to an external seminar at which they were taught methods for implementation of TQM. They were then given the brief to commence improvement activities at each store. This program faltered due to a lack of skills to deal with practical issues which occurred on a day-to-day basis. Safeway now also recognises that senior management were themselves not totally committed to the implementation process.

In mid-1991, nine area managers attended a conference on TQM in Melbourne. It was an educational experience for them, as the realisation dawned that much more was involved in TQM and its implementation than they had previously thought. They resolved to develop a strategy which highlighted the commitment by senior management, and which included comprehensive training in TQM for all employees.

While Safeway believe it is still in the learning stages, the strategy is now clearly being followed. The 'Safeway Team Participation' program (the name given to their TQM program) was developed internally, and information kits are distributed to all employees. The general manager devotes an average of one day per week to personally delivering lectures. He achieved a target of 40 lectures in 1992.

Barriers to implementation

Apart from problems in the early stages of TQM at Safeway several barriers to implementation of the Team Participation program remained. Understanding of the principles of TQM by store employees remained poor, and many managers tried to 'cheer it into the organisation', as one manager put it. Although Safeway employees have always enjoyed a good 'family' relationship, a strategy to address the differences in culture between the nine warehouses, 20 departments and 132 stores was required.

Safeway management addressed some of the issues by appointing 'sponsors' for each improvement project, to ensure that good suggestions were communicated throughout the organisation. The role of sponsors was clearly specified, including the responsibility to provide discipline and update senior management on the progress and continuing actions of each improvement team. Formal systems were also introduced to provide support after training, and the discipline necessary to ensure follow through of projects. These systems are discussed later in this case study.

Role of quality system

A formal quality system as outlined by the AS3900/ISO9000 series has not been fully adopted. However, some aspects of a formal quality system are evident. Although training manuals and videos for all operating jobs in warehouses and stores have been in existence for a number of years, not all have been regularly updated.

There are at least two methods which record errors: store complaint reports, and a 'circle line' system. It is a requirement for warehouses to always have ordered items in stock, so that stores can immediately be replenished on request. If a storage rack reserved for a circle line item happens to be empty when the stock is required for dispatch, a 'red circle' is placed on the rack. The red circle represents an out-of-stock product which must be rectified immediately. The number of red circles dropped from 3.5% in December 1991 to 0.1% in June 1992.

Complaint reports from stores are addressed by a Store Variance Report, which must be actioned by the state distribution manager and warehouse manager.

Early in 1993, Safeway began to review the applicability of the AS3900/ISO9000 Standards to their business. One senior manager expressed an opinion that it is more important for improvements to be directly visible. Hence he believes that Safeway's quality should be judged by their customers, rather than by a third party certification authority.

Defining quality in terms of customer perceptions

Safeway uses several methods for gathering customer information including market surveys, focus groups and suggestion boxes. Market surveys are conducted on a regular basis by professional market research organisations. Some stores have held focus group meetings to better understand customer wants at a more personal level. Safeway have found that, although they are not paid for these sessions, customers are very willing to participate in the meetings. The frequency of focus group meetings has increased recently with most stores now holding a focus group meeting during morning tea every four to six weeks.

When the focus group meetings first began, participants were asked whether they perceived any problems with the service they received. As the majority of responses to such questions were 'no', staff interviewers found it difficult to identify problem areas which could be addressed. Safeway interviewers have since learned to ask customers why they sometimes did not shop at Safeway for deli items, vegetables and other fresh foods. The responses to such questions have provided Safeway with useful information on how they could attract more customers.

Suggestion boxes have also been started in some stores and have proven to be quite successful. They are also used as a means for customers to report complaints. At one store, suggestions were received at a rate of 20 per week. Unfortunately, official figures for the percentage of suggestions which were implemented have not as yet been compiled. Customer complaints are recorded in a log book, however, and each is individually followed up by contacting the customer.

Safeway works hard at satisfying their customers. For example, one store maintains stock of a product which is not a shelf item, simply to meet the needs of one particular customer. The product is kept in a special room for the customer, who has continued to visit the store regularly for many years. In other instances, Safeway staff have specially purchased products from competitors to satisfy regular customers.

Three key indicators of customer penetration are used within the organisation:

- number of transactions per week
- average amount (in dollars) of each transaction
- sales breakdown by product category.

The data received from each store allows the store's performance to be monitored.

Information from market research indicates that Safeway is generally perceived to be a convenient, one-stop supermarket. The information also indicates that consumers perceive Safeway to be more expensive than its competitors. However, Safeway firmly believe that this consumer perception is unjustified, as their own information confirms that their prices are not dearer than their competitors' prices. Safeway attribute this consumer perception to two factors:

• cleaner and more presentable stores, which create an 'upmarket' atmosphere;
• the fact that stores are replenished nightly; consumers are more likely to believe that the higher wages for night work are passed on to customers.

Safeway tries to correct this consumer perception by continually advertising lower prices.

Continuous enhancement of processes

A formal structure (Figure 1) is used to monitor and control quality improvement activities throughout the organisation.

The Quality Councils are a three-tiered structure, originating at store level, progressing to area level, and finally to State level. These are designed to review team project results, and may authorise operational changes or expenditure. The Quality Councils also enable successful ideas to be communicated company-wide.

Although there are no franchises within the Safeway group, each store is managed as an individual business unit. Improvement projects are therefore managed predominantly by store managers, although the sponsors appointed for each project may be at a higher level in the organisation. Naturally, the benefits realised through continuous improvement activities vary from one store to the next.

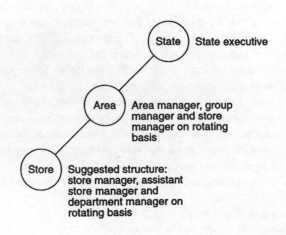

Figure 1   *Quality council structure*

Some of the more successful projects are described in Safeway's *Quality Development Bulletin*, issued on a monthly basis. Some excerpts from the bulletin are included below.

Issue No. 4, September 1992
Project Objective: To increase meat sales by $3500 per week in two months (Vermont store).

At first we all sat down to come up with some new ideas. The areas we looked at were getting meat out earlier in the morning, more time on the PA and more facings of bulk packs because this store is a high volume bulk pack store. Well, we moved the bulk packs out of the cabinet they were in because it was too small - it only has 52 facings. Now we have them in a cabinet with 96 facings. (Vermont's bulk packs are beef, lamb and pork). Then we started rearranging our work schedule. Now we start at 6:30 am in the mornings – a half hour earlier so that we can get the first trolley of meat ready to be put out by 7:30 am. Now customers who come in early have a full range to choose from. It took us a couple of weeks to fine tune it all but now it is working well. Even our late night customers have a much better selection. This is one of the areas where we were losing sales. When I started here we were running at about $28,000 per week. Now we've got that up to $35,000 per week. My next goal is to have a go at the department's all-time record for a normal week – $38,397.

Issue No. 3, August 1992
Project Objective: To increase scanning rate from an initial 12 items per minute to 15 items per minute by July 31, 1992 (Mt Waverley store).

In November 1991 we were told that the scanning rate had been set at 15 items per minute. So our Customer Relations Committee decided that would be our first goal – and then we would set about to better that. We immediately realised that our scanning checkstand design was new, upright instead of horizontal and we were told that this meant we should be able to get 17 items per minute. So that became our second goal. We actually got people scanning at 17 items per minute by the completion of our first goal on July 31. The store average has gone from an initial 12 items per minute to over 16 items per minute in the first 6 months. We have graphed our progress since January 1 and as you can see we are making steady progress, 80 per cent are now scanning at 16 items per minute – those that haven't are casuals who only work one day a week. The benefits of this effort have been an increase in productivity and customer service, and a reduction in the cost of doing business. Everybody is very keen to cooperate. Every Saturday we do a printout of our progress – the previous week's scanning rate. If we are a little slow in putting it up for inspection then I have the operators asking what the scanning rate is.

An example of another improvement which has been realised company-wide, but which has not been reported in the bulletin is a reduction in 'bad merchandise' (spoilt or damaged goods) from 1.29% of sales to 0.4% over a period of six weeks from July to August 1992. Some of the steps taken were the implementation of a 'repair shop' for bad merchandise, training of staff in handling of merchandise and introduction of a 'rotation card' which signals when stock should be reviewed for rotation.

Quality costs

Although costs within stores and the group are measured, there is no formal quality cost measuring system. This is an area which Safeway has not yet investigated.

Vendors as partners in the system

There are a total of eight buying divisions within the Safeway group. Some of the activities of these divisions overlap and, as a result, purchasing policies are very much left up to individual merchandising managers, within corporate guidelines.

Safeway has worked on improvement initiatives with some of their key suppliers to form 'win–win' alliances. In one instance, Safeway took an unconventional approach in their negotiations with a supplier of dairy products. The two parties agreed to a policy where mutually agreeable terms could be negotiated. In return, Safeway agreed to certain conditions put forward by the supplier. As a result of these open discussions, the supplier agreed to invest A$1 million in plant upgrades to provide a total service to Safeway, including packaging of the product.

In recalling this experience, the product buyer at Safeway explained that the breakthrough could only have been achieved through honesty, integrity and communication between the two parties. He firmly believes that whilst quality, cost and delivery are important, they are only the results of honesty, integrity and communication. In his opinion, alliances rather than deals with suppliers help to build loyalty between the two parties.

When prompted about maintaining preferred and approved suppliers, the buyer expressed a personal belief that such practices were contradictory to establishing loyalty between customers and suppliers, for 'approved' suppliers are generally viewed by companies only as backups if the 'preferred' suppliers temporarily fail to perform. In any case, most of the products sold at Safeway supermarkets cannot be substituted by other brands.

Safeway encourages its suppliers to participate in reviews of market performance and competitors. For example, the delicatessen department reviews market information with suppliers on a six monthly basis. Suppliers are invited to discuss issues such as market growth, sales of particular products, and performance against competitors. It is possible to involve suppliers in assessment of Safeway's competitors because each delicatessen product is generally sold through one supermarket chain only.

In addition to these reviews, Safeway has hosted a fair where suppliers were invited to present their products. Orders were not taken at this fair, as it was held solely for purposes of exchange of information.

Statistical thinking and methods

One of the aims of the Team Participation program is to promote the use of simple problem solving tools within the organisation. Interestingly, Safeway have opted to include only a few tools in their staff training. These include flowcharts, fishbone charts, line charts and brainstorming. There are two reasons for this choice:

- simplicity of approach, to cater for the diversity in cultural and educational backgrounds of their staff;
- lack of applicability of other tools to their operations.

Indeed, many problems are addressed at improvement meetings through discussion and sharing of ideas rather than through the formal use of problem solving tools. Line charts, called 'scoreboards', record the progress of each team. Variation and its effects is not included in staff training. Senior management feel that statistical control techniques have little application in warehousing and retail businesses.

Employee involvement

Safeway places particularly strong emphasis on employee involvement, for two possible reasons. The first is related to the industry; whereas much of the work in manufacturing organisations is task-oriented, the work in service organisations such as Safeway is highly relationship-oriented. Hence it follows that Safeway need to maintain a highly motivated workforce to provide superior customer service.

The second probable reason is the general manager's personal values. Clearly, employee satisfaction ranks high on his list of priorities; it was his approach to people which started the turnaround of an unprofitable store into a success (discussed earlier).

The general manager believes that four factors are necessary to ensure success in TQM – involvement, empowerment, measurement and rewards. An analysis of training materials reveals how these four factors are integrated in the Team Participation program.

Involvement

Every manager and chairperson of a team at Safeway is required to undertake training in the Team Participation program. Each store is required to form improvement teams so that employees are given the chance to be involved in an improvement project. However, it is not compulsory for employees to join a team. The structure of the team is clearly specified in the Team Participation Information Kit, along with the responsibilities of each team member. This formal structure helps to introduce discipline, commitment and continuity to the purpose of the team.

Measurement

At the start of each project, each team is required to identify the objective and select a quantifiable measure of progress. A rating system is used if the agreed measure cannot readily be expressed as a number, such as the degree of cleanliness of a staff room. The team is then provided with a large sheet, called a scoreboard, on which progress must be recorded regularly. Scoreboards and minutes of team meetings are prominently displayed in the office area or staff room.

Empowerment

People are encouraged to undertake improvements by themselves. Often, the *Quality Development Bulletin* reports on individual rather than team efforts. Checkout operators are empowered to make decisions on behalf of the store. For example, if the price of an item is not clearly marked on the item and the customer is in a hurry, the checkout operator has the authority to accept the customer's estimate of the price. If a regular customer has forgotten to bring his or her wallet, checkout operators also have the authority to accept the customer's word to pay for the purchases on the next day or visit.

Recognition and rewards

Safeway uses several ways to recognise the achievements of employees. For example, if a customer who is particularly pleased with a Safeway employee notifies head office of the employee's efforts, that employee receives an 'I Care' badge. This badge is presented by the general manager at the store, during a gathering of store employees.

The *Quality Development Bulletin*, which reports on achievements on a monthly basis, is distributed throughout the organisation. The name and photograph of the employee or team is featured prominently in the bulletin. A certificate of recognition is also presented by the store manager to employees who have achieved a significant improvement within the store.

Financial rewards are not generally offered as incentives for quality improvements, due to the difficulty of placing a financial value on the improvement actually achieved. There is also the risk that the wrong message may be conveyed to employees. Safeway often give away free hampers, celebrate employees' birthdays, and hold netball competitions. Such initiatives are aimed at creating a 'family' atmosphere within the organisation, and at building teamwork among employees.

Integration of continuous improvement activities with strategic planning cycle

The initial improvement projects at Safeway were chosen on an 'as-needed' basis, often by store employees who were not necessarily aware of the strategic aims of the company. Whilst these improvements contributed to overall efficiency in the organisation, a more systematic approach could help to ensure strategic goals are realised faster.

Late in 1992, Safeway management decided that four mandatory project teams were to be established within each store. These were:

- customer focus teams, to identify and satisfy customer expectations
- staff focus teams, to improve staff working conditions and morale
- 'cost of doing business' teams, to identify ways of reducing operating costs
- 'grocery bad merchandise and distress' teams, to reduce the level of spoilt and damaged grocery items.

Area managers are now also required to undertake self-appraisals of their own stores on a regular basis. The results of these initiatives have not yet been analysed.

Benefits of TQM at Safeway Australia

The general manager commented on the benefits which resulted from their adoption of the TQM philosophy. He believes that the key benefit has been a better understanding of customers' wants, which allows Safeway to provide superior service to customers. Some of the initiatives taken by various stores were:

- an umbrella service on wet days
- free tea and coffee in 'free-standing' stores (stores not located in a shopping centre)
- valet service at the affluent Melbourne suburb of Toorak
- entertainment for children at stores.

Some unexpected benefits have also resulted from the Team Participation program. The general manager is pleased to observe the significant improvements in team spirit, morale and personal development of employees since the program started. He also makes the point that employee loyalty and honesty affect the way customers behave. Goods lost through internal pilferage as well as shoplifting have reduced dramatically in those stores which successfully implemented Team Participation.

The general manager readily acknowledges that often, these benefits cannot be directly measured or traced to a particular initiative. However, he feels that some benefits have been so obvious and immediate that it is difficult to ignore the contribution of improved morale among employees. In one instance, a store reported a threefold increase in sales over 12 months due to a simple improvement. This store's employees signed a pact amongst themselves to greet each and every customer they met as they worked in the store. They also found far fewer empty food packets left on the floor, indicating a decrease in customer abuse of the system.

Although the Team Participation program continues to bring about change and improvements, its introduction was preceded by other successes through TQM. An illustration of the cultural differences which existed within the organisation was the warehouse situated in Bayswater (an eastern suburb of Melbourne), which had a long history of industrial relations disputes, in-fighting and unacceptable levels of waste and inefficiency. Safeway management were forced to choose between two inevitable options; either improve the work environment dramatically, or shut the warehouse down.

Fortunately, they decided to give the warehouse one last opportunity for change, and in 1990 appointed a new manager to oversee operations. His initial attempts to introduce a new work ethic through employee cooperation met with failure. In desperation, he decided to present the facts to a meeting of all employees, and firmly resolved to form improvement teams. Each team reported their progress and future plans directly to him on a weekly basis. Slowly but surely, some teams began to report improvements, which provided the impetus for change in attitudes of other employees. Today, the Bayswater warehouse is an efficient part of the Safeway organisation.

CEO'S comments

In a recent public address, the general manager gave the following comments on his experiences at Safeway:

The introduction of TQM here in Victoria has been a long, hard road – in fact, we're now close to four years and I can honestly say it has been one of the most rewarding experiences of my career, particularly now after all the hard work and frustration, I am starting to see area managers developing their store managers into teams to improve their operating performances. When I look around at the State now and see the improvement in customer services where stores are doing things for customers that I would never have thought of with our old traditional thinking about customer needs and services. However, we have certainly reached the point now where we have clear proof that the philosophy of TQM works and have the case studies available to show how it can improve sales in a store and bottom line profits are added through expense reduction programs.

Discussion

TQM is difficult to implement in organisations that have a large number of casual employees, who may not share common goals with others in the organisation. The task is made more difficult if employees come from largely differing backgrounds, cultures and educational levels. Furthermore, if a large number of sites are involved, the task becomes still more difficult.

Safeway, a Victorian supermarket chain employing over 17,000 people in 132 stores in the metropolitan and country areas, has managed to overcome these barriers in a fairly short period of time. Although the degree of TQM implementation varies between sites, it is still a remarkable achievement.

The factors that contributed to Safeway's success are quite obvious. The first is the public display of commitment from the general manager, to whom many within Safeway attribute their success. Sullivan (1992) writes:

It is not by accident that leadership is the first category in the Malcolm Baldrige National Quality Award (and Australian Quality Awards, 1993) criteria ... Only through personal, visible leadership can world-class customer satisfaction be achieved. A company's leaders must do more than talk about quality. Through specific actions – such as defining quality objectives, contacting customers regularly, routinely sharing customer information with employees, and emphasising customer satisfaction in goals and performance reviews – senior leaders promote a constant and consistent customer focus.

The effects of the general manager's actions clearly demonstrate Sullivan's point. Few managers with a similar level of authority make as many personal appearances, promoting quality throughout the organisation. This commitment from the top has clearly filtered down to the area managers at the next level of authority, who reinforce the need for quality in each of their stores.

Secondly, the subject of TQM has been simplified to a level where all employees can identify with the key principles. Only the simplest and most relevant problem-solving tools have been selected, and taught to employees in a simple and straightforward manner. These tools are prominently displayed on the walls of offices and staff rooms, thus reinforcing the need to use the tools for quality improvement.

The third factor is an effective system of recognition and reward. Safeway has clearly spent a great deal of time and effort putting together a system where individuals and teams can be recognised and rewarded for their achievements. This system clearly demonstrates that non-financial rewards can be at least as effective as financial incentives.

Carder and Clark (1992) agree with Safeway's cautious approach to financial rewards:

> When substantial cash awards become an established pattern, it signals two potential problems:
>
> 1. it suggests that several top priorities are competing for the employee's attention, so that a large cash award is required to control the employee's choice;
> 2. regular, large cash awards tend to be viewed by the recipients as part of the compensation structure, rather than as a mechanism for recognising support for key corporate values.

As the Safeway quality program gradually becomes more sophisticated, management will need to recognise the role of other imperatives of TQM. These include the effective maintenance of a well-documented quality assurance system, and more widespread use of statistical techniques to manage variation within the business. However, Safeway do not appear to be an organisation which takes things for granted. They are already market leaders in Victoria, and are realising better profits since adopting TQM.

Conclusion

Based on the Safeway experience and that of the other companies which we have studied, several factors are identified below which are more likely to contribute to the success of a TQM program. The experience amongst the organisations studied has been to use a particular methodology to introduce the quality improvement program and create awareness at the highest levels of management, but later to modify or even revamp the methodology to suit the organisation. This typically happens about a year into the program. Many of the managers interviewed commented that 'off-the-shelf' packages help to create awareness of the quality philosophy but fail to deliver real benefits.

The factors critical to TQM success include:

* *Identification of the strategic direction of the organisation.* For many of the organisations studied, the initial TQM Awareness Workshop was the first time that senior management seriously considered the company's mission statement and policies.

* *Determination of customer expectations and measurement of perceptions.* Understanding and measuring customer perceptions is vital for continuous improvements. This information must be communicated throughout the organisation.

* *Clearly defined, and agreed to by all, strategy for implementation of the TQM program.* This can help to identify potential barriers and an understanding of the organisation's capability to deal with those barriers.

- *Establishment of a formal structure for controlling, monitoring, and reporting improvement initiatives*. Companies which did not put in place an effective system for doing these experienced a lack of credibility and suffered from a lack of enthusiasm in the long term.

- *Implementation of cross-functional improvement teams and natural work teams*. In many organisations these teams provided healthy competitive spirit for continuous improvements.

- *Implementation of a formal quality assurance system*. This helps to establish discipline within the organisation and improves communication. It also provides a sense of achievement and confidence.

- *Employment of suitable external consultants*. This person can help to overcome the initial barriers to the acceptance of TQM, particularly amongst senior management.

Although all the organisations studied had achieved substantial benefits, there still remained many opportunities for further improvements. One area was in the use of statistical techniques. Typically organisations only used SPC charts to show trends rather than use the results to control processes. Knowledge of statistical methods was lacking amongst many managers and is an area in which much more can be learnt and applied.

Acknowledgment

The authors would like to thank the management and employees of Safeway who willingly shared their experiences of the TQM implementation with the researchers.

References

Carder, B and Clark J B, 'The Theory and Practice of Employee Recognition', *Quality Progress*, vol 25, no 12, 1992, pp 25–30.
Sullivan, R L, 'Inside the Baldridge Award Guidelines', *Quality Progress*, vol 25, no 6, 1992.

Case study questions

1 The case study discusses a number of factors critical to TQM success. What evidence of these can you find in Safeway?

2 Discuss the contributions that casual employees can make to improving quality and productivity in an organisation.

C2

Continuous Quality Improvements in a High-Technology Manufacturing Environment

The company: Varian Australia

Varian is an international, diversified, high-technology US-based corporation with sales in excess of US$1 billion annually. Employing a total of 9300 people, Varian's facilities and support offices are located across the United States and throughout the world. Organised around an array of core businesses, it produces sophisticated leading edge products for a variety of fields, including:

- microwave, power, and special-purpose electron tubes and devices for communications, industry, defence, and research;
- analytical instruments for science and industry;
- wafer fabrication equipment for the semiconductor industry;
- radiation systems for cancer therapy and non-destructive testing;
- vacuum equipment and leak detectors for industrial and scientific processes.

Varian Australia is a wholly owned subsidiary located in Melbourne, Australia. Employing about 400 people on site, it is a completely autonomous operation which designs, sources, manufactures and delivers optical spectroscopy instruments. Customers include those from the environmental industry, university and research laboratories, and industrial laboratories in the chemical industry. Over 90 per cent of its products are exported, mainly to Europe and the United States. The Melbourne site is now recognised as the lowest cost operator in the Varian Associates worldwide group.

Total quality management at Varian Australia

It is difficult to define a starting point for TQM in Varian Australia. It is an organisation which, for almost two decades, has focussed on quality in its products and manufacturing processes. Various manufacturing programs, including Just-in-Time (JIT) and Value Added Management (VAM), have helped to provide a professional approach to manufacturing. Several quality-specific programs were also tried over the past few years. The present approach to business, called Operational Excellence, is being adopted by Varian worldwide and has general consensus as the most effective to date. More focus is now placed on customers, training and time to market. Monthly reports must also follow the format prescribed by the Operational Excellence approach.

Unlike many other companies, TQM at Varian Australia did not originate from pressing external forces such as falling market share, or a directive from international headquarters. In the mid-1980s, senior management searched for a competitive strategy to lower defect levels and thus ensure survival and growth. A separate Quality Department was established as part of this process.

Several approaches to quality programs were tried in the initial stages. In the late 1980s, managers evaluated the Juran approach to quality with the help of a consultant who provided some training by means of a series of videos and Juran training material. However, much of the training was directed at solving problems and it was felt that these off-the-shelf training packages were not effective in delivering a change in culture throughout the organisation.

The second attempt at establishing TQM in Varian Australia was launched in 1990, through a program developed internally by Varian Associates. Entitled 'Quality Workshops', it was a program which was to be used by all Varian subsidiaries world-wide. The need for cross-functional Quality Improvement Teams (QITs) and Corrective Action Teams (CATs) was established, but the program lacked focus on the strategic needs of the business. Although the material covered a wide range of techniques and skills, it was not successful in providing the means to convert theory into action. Much of the focus was on increasing awareness and attention to quality, but there was inadequate planning and structure for implementation of continuous improvement activities at all levels of the organisation.

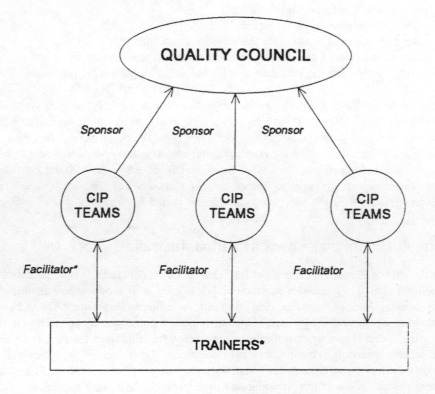

Figure 1   *Organisation which supports CIP training program*

The third and current endeavour, called Continuous Improvement Process (CIP), was started in late 1991 and is seen to be the most effective program to date. The CIP initiative has enabled a quality organisation to be established. This initiative has provided effective training, control and monitoring of quality improvement activities (Figure 1). Some strengths of the CIP approach are:

- focus on satisfying both external and internal customers
- driven by a formal Quality Council, rather than the Quality Manager
- formation of natural work teams as the principal vehicle for continuous improvement (as distinct from cross-functional teams whose specific objectives are to solve problems)
- emphasis on 'getting runs on the board' for small projects involving daily work
- 'train-the-trainer' concept, through to shop floor
- supports the goals of the worldwide Operational Excellence approach to business.

Implementation process

Varian sought the help of a local consultant to initiate the CIP program and deliver the initial training material. Senior managers who had previously received in-depth training in TQM were given a day of further training in CIP. Other staff were given approximately 20 hours of training each, and several staff have since been trained as trainers and facilitators in the program. In total, Varian Australia estimates that over 9000 hours have so far been spent on training and formal team work.

Varian began by identifying the specific needs to be addressed by the program. Key areas identified included team building, process improvement and problem solving skills. A Vision Setting Workshop was then held for senior managers, at which the strategy to achieve the goal of becoming the 'supplier of first choice' was discussed. The content of further training modules, strategies to meet Operational Excellence goals, and launch of the program were also planned.

Further courses were held to provide training for other managers and to facilitate implementation planning. One outcome of these activities was the appointment of a program manager to oversee CIP. Another outcome was the appointment of internal trainers to conduct further training within the organisation. Two pilot teams were then identified, one consisting of senior managers only. Each team selected an improvement project to manage.

Full implementation and training followed for the rest of the organisation. The first to be trained were indirect employees, followed by direct employees. Whilst being trained, teams also started on their individual improvement projects. Progress is reviewed on a quarterly basis by the program manager and trainers, using information from internal surveys.

Barriers to implementation

Because of its history of improvement initiatives, Varian's workforce was already accustomed to accepting new programs. However the experiences from previous attempts, which had not been extended to lower levels of the organisation, did little to create enthusiasm for CIP. The net effect was an attitude of disinterest on some people's part.

Other barriers included:

- perceived threat to supervisory and managerial roles;
- lack of credibility, as some good employee suggestions had previously been ignored.

Varian addressed these issues through several initiatives, as discussed throughout this case study.

Role of quality system

In November 1992, Varian Australia obtained certification to AS3901/ISO9001. Employees were asked if they believed the accreditation program had brought any benefits. In general, the answers were very positive. Although the decision to obtain formal third party accreditation was initially only a commercial need driven by changes in Europe, management now realise that it was necessary for the following reasons:

- quality system was previously ignored
- responsibilities were not sufficiently delegated, eg too many signatures were required to authorise a decision
- documents were not properly issued and controlled
- too many operational instructions were given verbally
- some testing equipment was not properly calibrated
- a systematic and logical design process was not followed.

Although all of these areas were subsequently addressed through implementation of the quality system, other benefits resulted. These included:

- better control of processes, resulting in consistency from design through to delivery
- increased measurement of performance, eg scrap rates
- a disciplined approach to business
- broadening of the need for TQM into other areas.

Managers were however unable to provide directly measurable benefits of the accreditation program, except that some government contracts would otherwise have been lost. Naturally, there were also criticisms of the quality system by some managers and engineers. For example, some felt that increased bureaucracy was introduced, and that the accreditation program became a goal in itself.

Defining quality in terms of customer perceptions

A key competitive strategy of Varian Australia is to continually market new and improved products. As a company dealing with high-technology equipment, product specifications are inevitably well defined. Specifications are developed from data obtained from several sources including the following:

- Customer Advisory Boards established in several countries
- surveys conducted worldwide with Varian sales force
- six monthly surveys conducted with selected customers
- innovative ideas from within the company
- competitive analysis.

Four key indicators which relate to customer wants are used by Varian:

- on-time delivery of complete systems
- quick delivery response
- product quality
- number of warranty incidents per unit.

To obtain this data, survey cards accompanying products sold are sent to customers. The response rate of approximately 50 per cent to these surveys in Australia is significantly higher than the response rate worldwide of approximately 10 per cent, possibly due to the absence of language barriers. Although the survey data is also reviewed at management meetings and by the Quality Council, it is seldom communicated throughout the company.

These surveys also revealed a number of positive and negative customer perceptions of Varian and its products. These include the following:

- Varian product technology is up-to-date
- the hardware is extremely reliable
- performance of the software used in Varian products is outdated
- Varian is not strong in product innovation.

To address some of these shortcomings, Varian Australia has begun to use Quality Function Deployment (QFD) as a design tool for all new products. One improvement which resulted directly from these surveys was an improvement in delivery response from ten weeks to four weeks. This improvement took three years to achieve and involved a complete revamp of existing systems.

Continuous enhancement of processes

Three major initiatives are currently being undertaken by Varian to improve processes – Just-in-Time/Value Added Management (JIT/VAM), Process Capability, and Value Managed Relationships (VMR).

Just-In-Time (JIT) and Value Added Management (VAM)

A JIT program has been ongoing for quite a number of years, overseen by the manufacturing manager. The VAM process was used as the principal vehicle to achieve its key objectives, which were to re-layout the whole factory, reduce set-up times, implement quality at source, and improve operations and systems. Kanban cards are used to pull materials through the line, and inventory is kept to a bare minimum. A materials store does not exist, as the majority of parts are unpacked and used directly on the production line.

To achieve confidence in the quality and delivery of materials, Varian initiated a Value Managed Relationships (VMR) program with its vendors in 1991. VMR is discussed later in this case study.

For the JIT system to operate smoothly, quality problems must be at a low level, and very much the exception rather than the rule. Quality performance is measured through a fault reporting system. Three levels of faults are identified:

- 'A' fault – out of performance or safety specifications
- 'B' fault – potential field service problem
- 'C' fault – cosmetic defect.

When an 'A' fault is reported, the 'A-team' (consisting of the production manager, manufacturing engineering manager, quality operations manager, and the supervisor or leading hand of that area) have to attend a line meeting regardless of their whereabouts at that moment, to solve the problem at its root cause. The line however, continues to run. Although it is a disruptive process, the frequency of 'A team' meetings has now been reduced to about once every two days, which is one-twentieth of its level five years ago.

Demerit points by product line are awarded according to the type of fault. A 'line stop' measurement system has also been established to record the number of instances when parts have not been available to the next process. The performance of each production department is measured in this way. Although cumulative graphs of both demerit points and line stops are prominently displayed in the production area, there were no obvious signs amongst employees that these graphs are viewed suspiciously. In fact, employees felt that the graphs help them to perform better through a healthy competitive spirit.

The results of the JIT/VAM program have been impressive:

- reduction of line shortage hours by a factor of three, within 18 months;
- halving of the number of people required, including inspectors, in the goods inwards area;
- reduction of factory cycle time (build and dispatch) from 20 to four working days over three years;
- greater than 95 per cent on-time delivery of complete systems;
- elimination of production control as a function.

Process capability

Varian Australia have recently embarked on a substantial program to introduce statistical measures of performance in the organisation. The Process Capability program is discussed later in this case study.

Quality costs

A Quality Cost system is in place which reports the following figures as a percentage of sales revenue for 1992: prevention costs, 1 per cent; appraisal costs, 1.5 per cent;

internal failure costs, 1 per cent; and external failure costs, 2.6 per cent (total quality costs, 6.1 per cent). Two years earlier, the total cost of quality was reported as 9 per cent of sales revenue. These figures are the lowest among Varian Associates manufacturing sites worldwide. However, the system does not capture waste costs associated with administration and service. For example, there are three to four employees servicing field problems and answering customer telephone calls. The cost of engineering changes made in order to rectify a recent design problem in one product was also not recorded.

Corrective Action Team (CATs)

Whereas the CIP program is primarily aimed at improving processes, Varian has another well-established system to attack product problems. Corrective Action Teams (CATs) were started in August 1986 in most production lines, and generally consist of employees working in the same area as well as some cross functional members. Team members meet once a week to discuss problems, as well as the implementation of other improvement initiatives to their particular area of work. Initially, the inputs to CAT meetings were internal problems, but the majority are now from worldwide field reports received through electronic mail. CATs have been instrumental in addressing product problems. Since their inception, CATs have resolved all but 200 of the more than 2000 problems identified (as of September 1992).

Varian believes the 'A team', CATS, and demerit points system have collectively played a significant role in making continuous improvement to products and operations. They have also allowed just-in-time performance to be achieved, and reduced warranty costs. With the CIP program extending to non-production areas, Varian is aiming to make continuous improvement a way of life throughout the organisation.

Vendors as partners in the system

Varian Australia conducts an active supplier program. Suppliers are listed in four categories as follows:

1. suppliers who bypass inspection
2. preferred suppliers for design of new products
3. suppliers on Value Managed Relationships (VMR) program
4. certified suppliers.

The category into which a particular supplier falls is determined by a system which rates their performance in terms of quality, cost, on-time delivery, communication, flexibility, documentation, packaging, and line stops caused at Varian.

An objective of the VMR program is to rationalise the number of vendors which supply the 5000 active parts to the factory, and thereafter to move towards sole supplier relationships. To realise this goal, several initiatives have been undertaken, including conducting seminars with vendors to introduce VMR, training selected vendors to reduce their set-up times, and implementing Kanban purchasing. Other measures

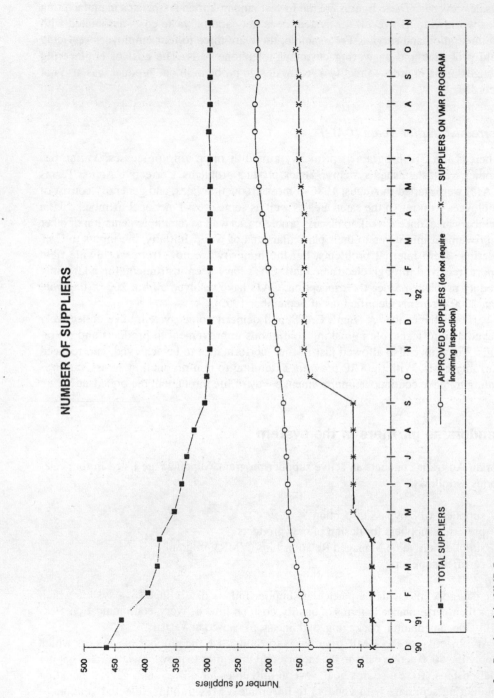

Figure 2   *Results of the VMR Program*

include the tracking of on-time delivery performance for all suppliers on the VMR program. Figure 2 shows the progress of VMR with respect to the number of vendors.

There was no available data on directly measurable benefits resulting from a specific vendor improvement project. However, overall pricing has been held steady over the last two years. VMR is also helping to build internal supplier–customer relationships within the organisation. Awareness is being provided through training in JIT/VAM and CIP, and relationships are strengthened through the line stop system.

It was obvious from interviews with employees throughout the organisation that the internal supplier–customer relationship is now established. Dramatic reductions in the number of line stops have resulted.

Statistical thinking and methods

Varian is an organisation which believes everything should be measured. A proliferation of graphs on display throughout the factory strengthens statistical thinking amongst employees. Approximately every two years, employees are retrained in the use of quality tools, so that it has now become a standard way of solving problems. Operators and leading hands alike are convinced of the power of quality tools; flow charts and CEDAC (Cause and Effect Diagram with the Addition of Cards) diagrams can be found in each area.

The insistence on data for decision making at management level has even caused 'analysis paralysis' at times, as one interviewee puts it. Opinion was divided as to whether all managers were themselves comfortable with the use of quality tools. The use of the 'seven tools of quality' is not always appropriate, however. For example, software bugs can often be traced to several lines of code quite readily. It follows then that the tools are useful only where a significant number of causal variables are involved.

Recognising the need to increase the use of statistical techniques, Varian enlisted Motorola's help in training and implementing a program to measure and reduce variation in the business. In 1992, almost 40 engineers were trained in Motorola's six sigma concept. The Varian program, entitled 'Process Capability', is being implemented as shown in Figure 3 and its progress is described below.

Firstly, the factory was divided into 10 main areas. Within each area, several key processes were selected and flowcharted. In the first year of the program, 68 processes had been identified and 44 flowcharted. Process capabilities were then measured and the results are now being fed back to engineering to assist in future design work. The flowcharts helped to clarify the points at which process control charts could be implemented; 15 control charts had been started by the end of the first year of the program. A benchmark is identified for each chart, and the team works towards meeting that benchmark.

What have been the benefits of introducing Statistical Process Control? Varian believe that it is still early days, but management is convinced that benefits will eventually flow. One key parameter in the ICPES (Inductively Coupled Plasma Emission Spectrometers) production line has already shown an improvement in Cp (capability index – number of times the natural variation in the process can fit into the

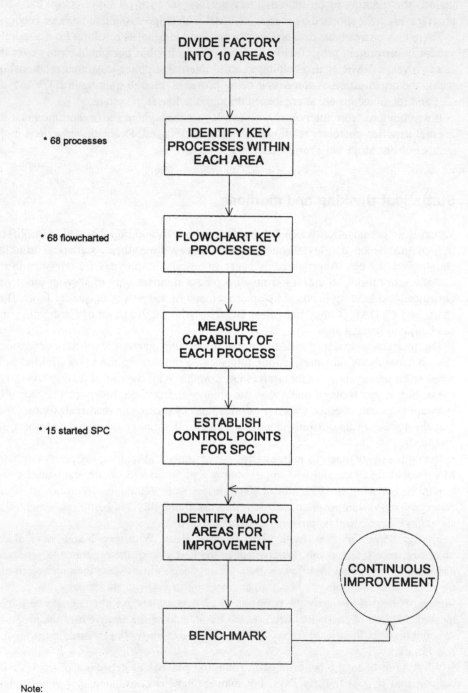

* 68 processes

* 68 flowcharted

* 15 started SPC

Note:
(* indicates December 1992 status)

Figure 3   *Process Capability program*

specification) from 0.7 to 4 since the program was introduced. The capability index is also beneficial as a tool to measure the 'goodness' of products other than by fraction defective, as defect levels continuously reduce. The focus is now on improving the weakest capabilities.

The use of statistical inference techniques in experimental design has not been introduced, nor are there plans in the near future for use of such techniques.

Employee involvement

It is now clear from the operators and leading hands interviewed that they truly believe in the aims of Operational Excellence. When asked why, responses such as the following were obtained:

> We can see that management are truly committed. If we ask for the right tools to do our job, we always get the tools, provided we can justify our request.

> The guys know what they are doing now and are able to run the shop. My role as a leading hand is only to keep production going, improve yields, and improve quality.

In the past, this leading hand's duties mainly involved expediting orders, but with the Kanban system in place it is no longer necessary for him to do so. How was this change in attitude brought about? One supervisor responded:

> I hate to use the cliche, but I think one main reason was empowerment. For example, we have given authority to certain qualified people in Production to raise Engineering Change Notes, without going through Design and Engineering.

People from the shop floor seem to support his belief, with statements like:

> We used to depend on QC inspectors to check the quality of our work. Now, we clearly understand that it is our job to produce quality; QC are there only on pre-production trials.

> We are all equals in our CAT meetings. The chairperson is really only there to take minutes.

Supervisors have benefited through closer cross-functional links with other departments and CATs, and an increased team spirit among themselves.

A number of methods to recognise achievement are used. These include monthly Operational Excellence newsletters, storyboards displayed on noticeboards for each team, and quarterly team presentations to the Quality Council. Internal quality perception surveys are conducted every six months. One problem which has been highlighted, and is being addressed by the Quality Council, is a lack of response to suggestions by operators.

Integration of continuous improvement activities with strategic planning cycle

The annual Strategic Plan forms the basis for Operating Plans which are formulated by each department. All Operating Plans include specific quality objectives which are given the highest priority for discussion in management meetings.

Varian have also employed some novel methods to ensure involvement in quality. Each manager is given a set of objectives for the next year. The overall rating scored at performance appraisals however, cannot exceed the rating scored for meeting quality objectives. This agreed system ensures that all managers focus on quality issues above all else.

Leadership in quality

During a personal interview with the managing director, Philip Thomas, several questions were asked. The following are his responses

TQM programs all too often fail through a lack of credibility after several attempts. What factors contributed to Varian's success despite several unsuccessful attempts?

I would not classify our previous attempts as unsuccessful, but more as parts of an evolving process. For many years our company has had a high priority on quality, and product safety has been paramount. However, I would say we were not organised for a total company-wide approach to quality until 1986.

The Varian Corporation had developed some training materials at that time but the majority of the material they used was provided by a US-based consulting organisation. We decided against using the same consultants in Australia because their nearest operation was based in Singapore, which was far from ideal. Furthermore, we felt that the consultant's approach was not culturally well suited to Australians, being too strong on motivational approaches such as 'Zero Defects Day' and quality rallies, and lacking in the 'how-to' of quality improvement.

We then selected an Australian consultant with whom the senior management team spent three days off-site. The material was the Juran approach to quality improvement. Unfortunately we realised that the consultant had very limited hands-on experience in implementing a quality program, and although the Juran material was excellent, we were going to have to develop our own approach. In 1986 there were few, if any, consulting firms in Australia skilled in implementing a quality program.

At the same time, our corporate quality organisation in the USA began to produce several quality training products such as Quality Improvement Workshops, Basic Quality Tools, and Supplier Relationships. We trained all our employees using a combination of the Juran material and our own material.

During the first five years of our quality initiative there were significant improvements achieved in many key indicators such as warranty costs and the production quality index. However, we still had a very product-orientated view of total quality – not unimportant, but not the complete story.

In 1990 Varian appointed a new chairman and Chief Executive Officer who changed all that. We now operate to five key Operational Excellence Initiatives which have taken the quality 'ethos' into all parts of the company. These key tenets are:

- Customer focus in all that we do
- An unbending commitment to quality
- Flexibility and responsiveness
- Rapid time to market for new products
- Organisational Excellence.

With these concepts, we developed a 10-module training course for all employees, designed not just to identify and fix problems, but to examine everything we do to seek continuous improvement. We currently have all our employees actively participating in about 55 Continuous Improvement Process teams. In addition, there are numerous Corrective Action Teams – multidisciplined teams devoted to solving problems which may be beyond the capability of the work unit of the CIP team.

Until recently, pay back on our investment has not been outstanding, but we did not expect it to be. CIP teams were encouraged to select relatively simple processes for their first CIP activity to help them understand the process, and get quick results for positive learning and motivational reasons before tackling processes and problems which might be more significant, but be more complex and take longer to solve.

Nevertheless, even in the early exercises there have been many worthwhile improvements. Where we go from here is an interesting issue. I see CIP as a permanent way of life for business. We recently adopted the Motorola six sigma approach to process capability, and currently have around 30 processes in manufacturing operation under this concept, with pleasing results. We need to do more 'foolproofing' to prevent errors at their source.

It's a long answer, but I wanted to emphasise the evolutionary nature of this process. Previous approaches were not failures – just not as effective as our present processes for achieving a company-wide commitment to Total Quality, or CIP, or Operational Excellence, or whatever name you choose.

There is a common perception that the term 'statistical techniques' means SPC, however, statistical data analysis and Design Of Experiments (DOE) are seldom practised in this country. Is there a role for such techniques at Varian, and why?

Our CIP approach does not rely on very sophisticated techniques such as DOE. We rely on Pareto analysis heavily, but instead of DOE we use CEDAC charts (Cause and Effect Diagrams with Addition of Cards). CEDACs rely on brainstorming to identify possible causes, and then uses simple statistical tools to investigate those factors judged by the CIP team to be the most likely cause. They may select the wrong cause – but the desired improvement will not be achieved in this case, and another cause will be selected and investigated. There is, obviously a strong reliance on experience and intuition, but it works well.

We have not used DOE in our normal CIP process – at least not rigorously. I suppose CEDAC charts are a form of DOE. Our engineers and scientists have used DOE frequently as a tool for multi-variable control in product design and performance, or in chemistry applications such as studies of interferences in various chemical analyses.

Varian has set itself an ambitious goal by implementing a number of significant improvement programs in parallel. What strategies were discussed in order to manage all these changes?

In 1986 we realised we had to change our organisational structure. Prior to that time, I had just one quality assurance professional reporting to me. The Quality Control department, as it was then called, reported to Manufacturing. The entire quality organisation was pulled together under one manager reporting to me. We established a Quality Council comprising all of my direct reporting senior managers, and the strategy for implementation of our quality improvement process was essentially developed through that council. Naturally, the Quality Manager played a key role in setting the direction.

Resources were a major issue – both in terms of the material and people. I've already indicated how we solved the problem of generating training material. We've been very fortunate on the people issue. The senior management team is very capable and experienced. We have a good mix of visionaries, high energy people who have the ability to get things done and good team players, and all have great stamina, and a very strong will to succeed. It's a good mix. My approach with senior managers is to give them a lot of freedom to implement their ideas – but as part of a team headed in an agreed direction, not independently.

Although our programs have had some overlap as TQM developed over the years, we now only have a few integrated programs running in parallel – CIP, Process Capability, Just-in-Time and CAT (Corrective Action Teams). Yes, it stretches us to (and sometimes beyond) reasonable limits – but what's the alterative? If we are to continue our excellent record of growth and profitability – these things are essential.

We have a stable, cooperative workforce. Despite the fact that we have six different unions represented here we've not had any major problems. I believe most people enjoy the challenges of doing things better, and we give employees recognition for their contributions in monetary and non-monetary ways.

Did all this happen by design?

I think it's been more an evolutionary process than a designed process. We have a vision of what we want to become, but it wasn't always clear what the pathway would be. We are thorough in selecting our employees – perhaps painstakingly so, and having capable, imaginative people makes this sort of evolutionary development possible.

In general, what role will TQM have in Australia in 10 years' time? Will a philosophy based on teamwork remain feasible in a world where organisations continue to shrink in staff numbers?

One of the consequences of 'downsizing' is that it is sometimes the weaker performers who leave, or perhaps those who find the new culture not to their liking. My personal view is that teamwork is, and will remain essential. CIP is a way of using teamwork to tap the potential in the whole workforce, so a smaller organisation can still find ways to do new things.

As for the future of TQM – I am not sure what lies beyond TQM, but certainly companies who don't continually strive to be best-in-class are going to be in trouble. For the rest of my working life, the drive to provide defect free products, and services that meet customer needs will be a major activity.

Conclusion

The implementation of TQM at Varian Australia has not been an easy task. Varian management have only recently found a quality program suitable for the company since beginning their search in 1986.

Some of the reasons why previous programs did not fully satisfy their expectations included off-the-shelf packages which did not contain business-specific information, inexperienced consultants, and programs which were strong on training but weak on bringing about change. Varian also contributed to this lack of progress by not having clearly identified goals and needs, and failing to follow up actively through improvement projects. Varian's current program (CIP) is seen to be different because the needs of the business were identified up front (through the Operational Excellence goals), and improvement activities were implemented throughout the organisation, including non-production areas. Moreover, it concentrates on improving processes as well as solving problems.

Together with CIP, Varian had to implement several major programs such as JIT and VMR. According to the managing director of Varian Australia, effective organisation, as well as the ability and teamwork of senior managers, were the keys to effective management of these overlapping projects.

Although enthusiasm about CIP is high, it remains to be seen whether it will deliver the benefits expected. Responses from internal surveys indicate that some perceive the organisation as slow to react to employee suggestions. Much training has been spent on statistical techniques, and the results of these efforts may not be known for some time.

In some ways, Varian Australia are still only beginning to adopt Total Quality Management on a company-wide basis. But then again, they appear to be an organisation that is willing to wait for benefits in the long term.

Acknowledgment

The authors would like to thank the management and employees of Varian Australia who willingly shared their experiences of the TQM implementation program with the researchers.

Case study questions

1 Discuss the role of senior managment at Varian Australia in the various improvement initiatives leading to the current CIP program.

2 Discuss the application of statistical methods at Varian Australia. Do you think it is important for shop-floor employees to understand and use CEDAC and DOE techniques?

3 Comment on Varian's JIT and VMR programs and their relationship to the quality improvement initiatives.

C3

Quality Improvement at Van Leer Australia

The company: Van Leer Australia

Van Leer Australia is a subsidiary of a multinational corporation involved in the manufacture of food packaging, plastic containers and steel drums. The company employs a total of 15,000 employees worldwide, with its headquarters situated in Amstelveen, Holland.

The Australian manufacturing plant is located at Preston (Victoria) whilst the sales and administration office is located at Doncaster (Victoria), both employing a total of approximately 120 people. Van Leer Australia manufactures six products in Victoria, including fruit trays, paper plates, egg cartons, plastic meat trays and plastic service trays. The company supplies products in high volumes to customers in the supermarket, farming and meat industries. The critical market success factors are cost, product reliability, and delivery performance. External threats to the company include lower priced products manufactured locally, imported products from Asia and New Zealand, as well as environmental issues concerning the use of plastic in products. Management recognises that to compete effectively, the company must focus on improving plant efficiency, reduce inventory levels and increase autonomy among employees.

The Preston plant employs a multicultural workforce. In general, production operators are not required to have specialist skills to be considered for employment. Some of the production operators are also not fluent in English. Although this does not usually pose a problem in daily activities, it can be difficult to communicate new initiatives to all levels of the organisation.

This case study is based on interviews conducted with eight Van Leer Australia employees from marketing, production, and process engineering functions. The levels of responsibility of these people ranged from general manager to production operator. Much of the detailed information was provided by the quality system coordinator, Mr Tony Pasalskyj, who had responsibility for maintaining the quality system, and continuous improvement activities.

TQM at Van Leer International

Van Leer was first introduced to TQM in the early 1980s when Dr Koos Andriessen, then Chairman of Van Leer International, led a corporate management team to a presentation in the United States by Dr Deming. They were impressed by Dr Deming's approach to customer focus, statistical process control and employee involvement. On their return to Holland, the management team resolved to establish a corporate philosophy of TQM, and to implement this philosophy within each of their sites worldwide. The name given to the Van Leer quality program was Quartet, an acronym for 'Quality And Reliability Through Expertise And Teamwork'.

The timeframe identified by the corporate management team for worldwide implementation of Quartet was two years. Initially, videotapes which described the Quartet program were distributed to each site. Training managers were sent to Holland for training in the program so that appropriate training courses could be established in their individual sites.

Unfortunately however, this approach lacked the emphasis necessary for effective implementation of the Quartet philosophy. Employees did not perceive the program as the corporate vision for quality, but rather as a course delivered by the training manager. Training was not provided to employees at lower levels, and any plans to implement Quartet were not clearly defined. A proliferation of Quality Control Circle projects which were neither documented nor followed up contributed to a lack of enthusiasm amongst employees.

TQM at Van Leer Australia

Quartet was introduced to Van Leer Australia in the mid-1980s when senior management introduced quality through an evolutionary process by continually stressing its philosophy. Projects which brought directly measurable benefits to the company were selected, and employees at almost every level were given a chance to contribute. As improvements began to happen, employees were increasingly encouraged by the results they achieved.

Table 1   Six crucial points from the Quartet philosophy

1 . The erroneous idea that there is a 'trade off relationship' between quality and quantity must be discarded. In fact, when proper attention is paid to quality, productivity improvement follows automatically.
2. The responsibility of those who work 'in' the system should be increased. They should be trained in the seven basic statistical tools; and properly trained to carry out their jobs. Above all, they must be trusted to do their job properly when they are given the chance.
3. The main task of those in management is to constantly work on the systems for which they are responsible. They are the only people with the power to change the systems. To do this, people must be trained in the use of the statistical techniques to provide the basis for continuous improvement. 'Imagineering' should also be stimulated. This is simply imagining how things would be if everything was perfect and comparing that with reality. In this way, it becomes easier to identify problems and propose solutions.
4. All levels of management, from the highest down, must participate and be totally committed to the 'new way of life' and give it time to become effective.
5. Outdated habits such as mass inspection (this is too expensive and occurs too late), and holding the workers responsible for all errors (in fact, 85% of the causes of unsatisfactory work, or quality, are caused by faults of or in the system itself must be discarded. It is management's job to change and improve the system
6. We must get it right the first time.

Senior management believe that, through this 'slow and steady' approach, Quartet is finally becoming a way of life within the organisation. For example, every meeting at senior level includes Quartet as an agenda item, and employee unions have also agreed to include Quartet objectives in position descriptions.

To increase awareness, a small handbook is given to employees as part of their training. The handbook contains the six crucial points of the Quartet philosophy (see Table 1), an introduction to the seven quality tools, and a systematic process for problem solving.

The Quartet program enables management of each site to measure progress and set future improvement objectives. This is achieved by the Quartet Matrix which identifies the 15 operating principles of Quartet and provides guidelines for measuring the degree of implementation of each principle. Management of each site are required to update the matrix on a quarterly basis by nominating the degree to which each principle has been implemented within their organisation, and the target for the following year. This matrix not only provides the focus for improvement activities within the site, but also a uniform measurement system for all Van Leer sites worldwide.

Role of quality system

Van Leer Australia achieved certification to AS3902/ISO9002 (Quality Systems for Design/Development, Production, Installation and Servicing) in April 1991. Employees at various levels in the organisation were asked if they believed the certification program had brought any benefits. Some of the responses were:

> The project may have been the wedge into union resistance, as they were involved in the project. It was the first sign of long term commitment to TQM by the unions.

> It has been beneficial, because people now have an improved interest in quality, and a better appreciation. However, we had to overcome linguistic problems in a multicultural workforce, as well as a lack of interest among some of the older workers.

> The ISO9002 project was fantastic. I changed my attitude to it when the processes were stabilised, and we did not have to touch the machines that often.

A comment made by several managers as well as shopfloor employees related to the introduction of bureaucracy into the system, because of the requirements of the Standard. One manager said they continually had to remind themselves to bring the Standard back to the real working environment.

Non-conformance and corrective action reports are used to capture and correct errors at the root cause. Van Leer initially elected to use these reports to cover all areas including complaints, production and administration. However, after a flood of reports were generated, some of a trivial nature, a review of the scope was clearly needed. The reports now cover problems of a non-routine nature which will not be solved by another system of procedure. At present, one or two reports are received by the quality system coordinator every month, and each is individually assessed.

When asked if directly measurable benefits resulted from certification, managers were unable to produce evidence of immediate benefits. The benefits are believed to be indirect and realisable only in the long term. However, at least two improvements to the system were identified – a customer survey, and a skills audit which resulted in formal training plans for employees.

One manager elaborated on two popular misconceptions about certification to the Standards. In Van Leer's experience, certification has not resulted in any new sales, only in internal improvements. He also does not believe that a formal quality system is necessary to ensure success in TQM.

Defining quality in terms of customer perceptions

Data on customer expectations and perceptions is obtained in three ways – individual sales contacts with customers, focus groups and customer surveys. The surveys were a direct result of the quality system certification program, during which a need for quantitative customer feedback was identified. Customer surveys are now conducted every six months with 10–12 key customers, and focus on three areas – perception of Van Leer as a company, product quality, and service.

The survey results indicate that customers generally regard Van Leer favourably with respect to its competitors. However, one manager expressed concern that these surveys do not necessarily indicate how the company performs in absolute terms, particularly if competitors also perform poorly. The results also indicated a perception that the company lacked innovation in its products, an area which is being addressed through better product design.

Performance data from surveys and other contacts with customers are presently communicated throughout the organisation through the Quartet Matrix. However, it is recognised that this method is inadequate, and management are experimenting with other methods of communicating customer expectations and perceptions to lower levels of the organisation. One method was to involve operators in visits to key customers. The responses from people involved in these visits were very positive. Although the visits have since stopped due to a lack of resources, Van Leer management recognise the benefits and are hoping to re-start the program.

Continuous enhancement of processes

A Quartet organisation structure has been established to plan and monitor improvement activities. The Steering Committee consists of senior managers, but improvement teams may have a mix of employees from all levels of the organisation. Each team has a team leader, who is nominated by the Steering Committee for the duration of the project. As a matter of policy, the team leader is given the authority to implement team recommendations, within the scope of the project. In general, these teams are cross-functional because of the project-driven goal. However, operators are represented by their supervisor because of the difficulties involved in holding team meetings during production shifts.

Some significant improvement projects have been successfully completed through this simple structure. Two examples are presented below.

1. *Printer improvement.* The objective was to reduce the set-up time required to change the printing machine from the manufacture of one product to another. This project was selected as a result of a Pareto analysis which showed that customers wanted smaller batch sizes. The Single Minute Exchange of Dies (SMED) concept was employed as the principal vehicle for improvement. Over a period of six months, the team reduced the set-up time from 80 minutes to 13 minutes. The next target is to reduce this to less than ten minutes, and to make the set-up procedure easier to manage.
2. *Reduction in chemical usage.* An estimated $130,000 worth of chemicals and dyes was lost annually due to wastage. A group was established to reduce wastage by 50% in the first year. On investigation, the team identified that 62% of the loss was from two chemicals. Two issues had to be addressed: whether the amount of loss was a true figure, and the means to measure the flowrate and usage.

   An action plan was instituted, which involved:

   • a procedure to ensure the correct quantity of chemicals are delivered
   • establishing control charts which monitor the use of chemicals during processing
   • ensuring that drums are emptied of all chemicals.

   These actions resulted in a reduction of wastage by 75%. The team continues to look for improved methods to handle chemicals and dyes, and are investigating the use of magnetic flow meters to monitor usage.

Management feel that financial measurements are but one method of expressing achievements. Furthermore, achievements may be difficult to quantify in financial terms. As a result, the payback of an improvement project is often measured in non-financial terms.

There is currently no incentive scheme to reward achievements by improvement teams. Results of projects are reported through several media – notice boards, the staff magazine and internal newsletters. Projects which have contributed significant benefits are submitted to the Van Leer International newsletter for publication. However, management has realised that more needs to be done to recognise the achievements of improvement teams, and are reviewing reward and recognition systems.

Vendors as partners in the system

Van Leer Australia maintains a list of approved vendors, an outcome of the quality system certification program. The company is further trying to establish sole relationships with some of its vendors by working with them on agreed improvement projects.

One example was a project undertaken with a major supplier of raw materials, with whom the existing contract had to be reviewed to meet Van Leer's requirements. Representatives of the supplier spent a full day at the plant in Preston to understand Van Leer's needs. In return, Van Leer personnel visited the supplier's plant to understand the production technology and constraints. The exercise clearly indicated a lack of capability on the supplier's part to meet Van Leer's production requirements. However, instead of resorting to other sources, Van Leer worked with the supplier to arrange an agreement which would benefit both parties. This resulted in a change of attitude from one of apathy to concern, and a commitment by the supplier to invest over $1 million in plant upgrades.

Through Quartet, management is also promoting awareness amongst employees of the need for internal customer–supplier relationships. However, a mechanism or measurement system to strengthen this philosophy has not yet been established within the organisation.

Statistical thinking and methods

Training in the application of the seven quality tools has been provided to every employee. At present, these tools are generally used by project teams for problem solving. It remains for management to ensure that the use of quality tools is accepted as a way of thinking at all levels within the organisation.

Van Leer have begun to analyse and monitor some processes using statistical methods. Attribute control charts have been set up to measure defect levels of several key processes. At present, the data collected is only used to calculate process means and control limits. Several issues need to be addressed however, before formal statistical process control (SPC) can be implemented:

- understanding of SPC principles by all concerned, to ensure data is accurately collected and interpreted
- commitment to controlling processes through capability limits, rather than specification limits
- stable and capable processes, to minimise line stops and other disruptions.

A plan to address these issues is currently being developed. SPC will gradually be implemented to include all relevant production processes. There are no plans at this stage to introduce statistical inference techniques or Quality Function Deployment (QFD) to the organisation.

What will be the benefits of introducing SPC? Van Leer believe that although it is still early days, SPC will provide the needed emphasis on measurement of process performance. For example, the number of defective units produced by the egg tray production line is not currently measured. Because all paper waste can be recycled, there has been no emphasis in the past to reduce waste on this line. Indeed, the waste produced by most of the production processes can be recycled. Through SPC, the levels of waste in each process could be reduced.

Employee involvement

When asked about their reaction to Quartet during its introduction to Van Leer Australia, employees at various levels in the organisation responded with comments such as:

'My initial reaction was that it would be more work for us.'

'I reacted negatively when the posters started going up all over the place.'

'My first perception was that it was not backed by management.'

Six years later however, these same employees have clearly changed their attitudes. One of the respondents became a strong supporter of the Quartet process, after seeing for himself the results of the improvement project which he had led. Indeed, similar reasons were given by many respondents who had changed their attitude about Quartet.

According to one operator, the quality of products has improved tremendously since he first joined the organisation four years ago. He attributes this improvement to better training.

'When I first joined, no-one knew how to operate the machines properly. The amount of waste in this place was incredible' he said, pointing to the production area. 'Now, blokes try to make the job easier for each other, and they know how to set the machines and make them run properly. Look at the place now – there is no waste anywhere.'

A statement made with such pride would lift the spirits of management in any organisation.

Integration of continuous improvement activities with strategic planning cycle

Improvement projects at Van Leer Australia are currently chosen based on customer feedback, and internal opportunities for improvement. The Quartet Matrix plays a major role in identifying and monitoring the progress of projects. The Quartet philosophy does not provide for the integration of improvement activities with the annual planning cycle. However, management have already recognised the need to work towards this aim, and plan to integrate Quartet initiatives with future business strategies.

Discussion with Van Leer's CEO

During a personal interview with the General Manager, Mr Martin Wood, several questions were asked. The following were his responses.

What were the factors which contributed to the decision to adopt a 'slow and steady' approach to quality, rather than a high visibility improvement program?

I think it was probably Van Leer International's corporate culture. It is a 'steady' company, not a high-flying company. The company has always been run on conservative balance sheets; I think the people at Van Leer have been brought up that way.

Do you think a steady or deliberate style of management is appropriate today?

No, I think we need to speed up. The Chairman is trying to change the corporate culture. But remember, Van Leer has been doing TQM for 10 years – how many companies have been at it for that long? It's almost necessary to take that long. I also think that TQM may be stronger in our European operations than in Australia, since these operations are closer to where Quartet was developed.

What are your perceptions on quality programs which have been developed overseas for local implementation? Was there any resentment, and would it have made any difference if the program had been developed locally?

I think they are trans-national, and should be that way. There was some resentment, but I also think it will always be that way – 'another Head Office initiative!'. It would not have made any difference to attitudes if it had been developed here.

In retrospect, what would you do differently if you had to implement Quartet again?

If the main instigators practise it themselves, then the change would have been faster. If I had any criticism, it would be that. How do you think people would feel if they were told 'You will do Quartet; it will be done this way'? These sorts of things often happen that way – it wasn't sold properly.

In general, what role will TQM have in Australia in ten years' time? Will a philosophy based on teamwork remain feasible in a world where organisations continue to shrink in staff numbers?

We have restructured a lot recently. In fact, because we have less people, it has been easier to implement TQM. We are becoming better at it daily because of restructuring. There is better communication now between the remaining levels.

   With regards to the future of TQM, I think it is a very difficult question to answer. The world is changing so quickly. I haven't personally seen any new philosophies on the way.

Conclusion

The decision by corporate management to implement Quartet on a worldwide basis within two years may have been an overly ambitious goal. Berry (1991) observes that, 'complex change takes years and will never occur as a result of management pronouncement'. However, other observers feel it is possible for a company to turn around within six or eight months of working with a (consulting) firm (Crosby, 1991).

It may be difficult to evaluate the success of this two-year plan, as the original corporate objectives have been lost through time. What is apparent however, is that the change in employee beliefs did not occur as a direct result of management efforts to address negative attitudes. Rather, it resulted from the 'slow and steady' approach used by management to achieve results through employee involvement. Although the change was gradual, it has clearly been effective in the long term. Apart from a few Quartet matrices and other references to AS3902/ISO9002, 'Quality' propaganda is not highly visible in the working environment. This agrees with observers who believe that in future, quality will not be limited to quality specialists, but will be an integral part of what managers do for a living (Axland, 1993).

One problem with this low profile approach is the differences in implementation methodology among various functions. For example, one manager does not believe that the formal structures prescribed by Quartet is necessary for developing teamwork or for solving problems. Another manager believes that documenting each project may sometimes be detrimental to employee initiatives.

Despite any barriers or disagreements the company may have faced in the past, one thing remains clear. Van Leer Australia has continued to survive and improve after ten years of TQM and Quartet, in a period of worldwide recession and an environment of stiff competition.

References

Axland, S, 'Forecasting the Future of Quality', *Quality Progress*, February 1993, pp 21–25.
Berry, T H, *Managing the Total Quality Transformation*, McGraw Hill, 1991, p 29.
Crosby Associates Australasia Pty Ltd, *Survey on Quality Management; Interim Summary of Results*, Sydney.

Acknowledgment

The authors would like to thank the management and employees of Van Leer Australia who willingly shared their experiences with the authors.

Case study questions

1 Discuss the effectiveness of the Quartet program at Van Leer Australia in improving employee involvement and in changing the culture of the organisation.

2 Discuss the role of corporate headquarters in implementing TQM into an overseas subsidiary.

C4

Quality Management in the Smaller Company: Nally (WA)

Introduction

This case study is based on research which examined TQM at Nally (WA). The research involved various methodologies including:

- examination of company material
- interviews with company personnel
- interviews with members of Quality Improvement Teams
- observation of quality improvement team meetings.

A number of interviews were conducted with key persons involved in past and present activities. This included relevant managers, supervisors and plant operators. The main research objectives were to:

- explore the reasons for introducing TQM
- to examine the change process and examine the difficulties experienced in introducing TQM
- examine how TQM is presently operating in Nally (WA)
- evaluate the effects of changes brought about by TQM on employees management, suppliers and customers
- assess the overall impact of TQM on the company.

Company background

Nally (WA) is a small-sized organisation with approximately 60 employees. It produces a range of plastic products (mainly containers) for both the Australian and overseas markets using modern technology. Production for most of the year involves three shifts per day over a five-day week. In periods of high demand this may be increased to seven days per week.

The present company originally commenced operation in a new factory in 1985 and operated under the name of Viscount Plastics. In 1987, this company was bought by a Western Australian firm, Bristile Plastics. The new general manager, who had a strong personal interest in and previous exposure to quality management and teamwork was able to select a new management team who were supportive of TQM. In mid-1992, Bristile Plastics was acquired by a large international company, Pacific BBA (UK ownership 55 per cent, Australian 45 per cent), and now operates under the name of Nally (WA).

The workforce consists of a core of permanent employees and a number of casual staff. Core employees occupy mainly administrative, maintenance and supervisory

269

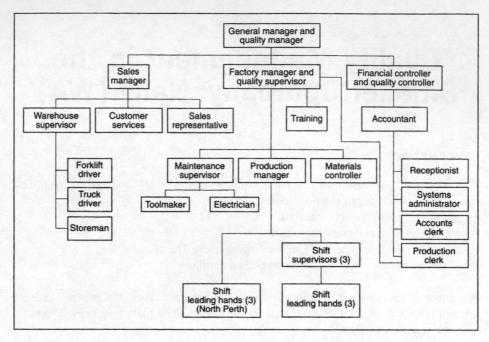

Figure 1   *Organisation Chart: Nally (WA) Pty Ltd*

positions. Casual staff are predominantly factory process workers. Union presence at the plant is not particularly strong with the Miscellaneous Workers Union representing shop floor operators. The structure of the company is shown in Figure 1.

Why TQM?

Following the company takeover the new general manager felt the need to develop a new culture particularly due to the large degree of mistrust and uncertainty between the former Bristile employees and the largely ex-Viscount management. TQM was also based on management's desire to return the company to profitability (both companies had been making losses at the time). This would focus on high quality service, use of sophisticated technology and the desire to be market leaders. TQM was seen as a way of helping to achieve these goals and a way of developing a team attitude within the company.

Soon after the merger an off-site three-day residential program for all managerial employees was held to focus on the need to change and be able to respond to continuous change. Several team building workshops which were designed to help develop cooperation in the new company were also conducted.

A memo from the general manager to all staff signalled the formal commencement of TQM. The company utilised funding which was available under the government NIES (National Industry Extension Service) program to employ a consultant for a training program. This consisted of a two-day course in TQM philosophy and techniques for group leaders and the steering committee (or TQM council), and a one-day course in TQM awareness for all employees.

Following this initial training, Bristile management planned the introduction of TQM throughout the entire organisation. Having only 60 employees made this possible to achieve within a relatively short period of time. The initial TQM structure consisted of a council and a number of quality improvement teams. The council (or steering committee) had overall responsibility for managing TQM and comprised the general, sales, administration, factory and production managers.

TQM structure and operation

Initially, eight quality improvement teams representing the major areas of the company were established. Some of the first team improvement projects dealt with reject and waste improvement and development of product specification. Since raw plastic pellets comprise about 45 per cent of production costs, early projects emphasised reductions in reject levels, general waste and the development of specific quality specifications. Other projects included better management of inventory levels and improving the cash flow.

One of the initial teams conducted a survey of employee morale. Lack of recognition was the highest scoring factor as a major contributor to poor employee morale. The survey results also pointed to a lack of procedures which affected morale and included; after hours despatch, emergency procedures, preventative maintenance scheduling, office hours, purchasing responsibilities, manufacturing hours, overtime procedure, telephone answering and car parking. Most of these pointed to the fact that the company had relatively few policies and procedures in place. Machine breakdown was identified by the maintenance group who later developed procedures for all machines. More recent issues dealt with by this team included; safety, performance appraisals, job descriptions, factory safety and even piped music in the factory.

Apart from examining one particular problem or improvement opportunity, several teams also discussed smaller problems referred to as 'leaky taps'. Leaky taps referred to small issues which could be resolved relatively quickly and often by one person. All team meetings still include these on their agenda. A noticeboard was placed in the factory meals room which listed 'leaky taps' and whether they had been resolved.

Some teams floundered, possibly as a result of the lack of extensive team leadership training and the generally limited overall training. In some groups, participation rates were very low. Several displayed a downward trend in attendance, whilst for others, attendance was relatively stable. The workload for team projects often fell into the hands of a few people who completed the work in their own time at home. The manufacturing teams were scheduled to meet prior to shifts which also presented logistical difficulties. In general, expectations were too high at the start.

Several changes were introduced in order to keep the process of TQM going. An outside consultant was used and members of the quality council attended team meetings to assist the team leader. A number of staff attended a facilitators training program. One of these now acts as facilitator for the manufacturing team. A quality policy statement for the company was developed and is shown in the Appendix.

During the same year the six manufacturing teams were merged into three as there was a degree of membership overlap. Further rationalisation of team structures occurred in the following year. The environment team was abandoned and the three manufacturing teams along with maintenance were combined. Reasons for the latter included a degree of overlap (the teams were differentiated only on the basis of the particular production shift). Attendance was also a problem despite the fact that participants were paid overtime for attending meetings. The administration and sales teams were also combined.

All company employees are invited to belong to one of these teams. The revised manufacturing team has met on Saturday mornings with wages employees being paid overtime for attending. Attendance is voluntary and approximately half the company's employees attend along with a number of administrative staff. Despite some initial reservations about how effective this single large (usually 30-plus attendees) team may have been, after several meetings it appears to be working well. A typical agenda is shown below.

Manufacturing Team Meeting Agenda

- Minutes.
- Report Team A. Ongoing process improvement tasks using data gathering, run charts, etc.
- Report Team B.
- Production manager's briefing on technical matters and processes. Fifteen minutes devoted to explaining terms, processes and equipment to team members as a general training exercise.
- New issues
- General manager's briefing or update on developments.

The meeting then serves as a general forum for communication on a range of issues as well as a process improvement activity.

All divisional managers participate in the meeting. Initial reaction from factory operators was somewhat reserved due to the involvement of the managers, however now their participation does not appear to have any inhibiting effect. On the contrary, many employees indicated that having the managers present is a plus since communication is immediate when issues and problems are raised at the meetings. This may be a very positive method of operating in a smaller organisation. It also helps to demonstrate management commitment particularly since they are giving up their Saturday morning, without pay.

An initial project for the revamped production meeting was selected by management and dealt with waste plastic. TQM tools such as bar charts showing weights and values of various types of waste plastic, cause and effect diagrams and brainstorming have been used. Being a large group, members from different shifts were allocated data collection responsibilities. Thus, for operational purposes, the large group is subdivided into several smaller teams.

Team meetings in the non-manufacturing area usually run along the lines of a departmental meeting where a number of issues are dealt with. Teams do not concern themselves with simply one problem or task statement at a time as the 'textbook' model might suggest. Instead they examine a range of issues and usually assign an action to a person or persons to cover outside the team. This is then investigated, sometimes introduced and reported back to the next meeting. A considerable degree of empowerment is provided for employees in the administrative area to actually introduce changes. The seven tools are not always used. Nevertheless, there is still a focus on continuous improvement by improving, simplifying or even developing processes. A further point is that given the cross-functional nature of most of these teams, for example the administration team, communication throughout the company is enhanced.

In the initial phase of introducing TQM, the general manager participated in all groups primarily to assist as a facilitator, although as teams acquired confidence they no longer required his involvement. Today, he is only involved as a participant in the administrative and production teams.

Problems and issues

As any organisation which has introduced TQM has discovered, problems and difficulties are encountered. What is noticeable in this company is the flexibility and determination to try and make it work by using new and alternative procedures such as new team structures.

An issue to surface relatively early was management's over-ambitious expectations of the teams. This was perhaps due somewhat to the limited training which had been provided, particularly in areas like leadership, team building and meeting skills. Furthermore, many employees had a relatively limited understanding of the equipment which they were using, making it difficult to identify problems. Participation in teams was also low. To address these issues, an external part-time facilitator was employed, further training provided, an induction program introduced and increased council support for team leaders provided. The production of a videotape for use in employee induction was also considered.

A variety of other issues which required attention included: overcoming the belief that management solves problems even though employees are aware of them, the feeling that TQM may be a fad and the fear that speaking up about problems at a team meeting could jeopardise one's job. To some extent this is still a problem which management and team leaders need to deal with. One problem with a company of this size and having several production shifts, is that it cannot afford too much time for employees to be away on training courses. Furthermore, the logistics of organising team meetings is an important issue.

Some felt that inadequate training had taken place initially, particularly for leading hands and supervisors concerning their role in the TQM process. The generally limited amount of training may have hampered TQM although some people do not respond well to any amount of training. It has been recognised that people and team building skills are important. It is also apparent that supervisors, team leaders and leading hands should be well trained prior to embarking on the TQM journey since if they are not well acquainted then they will be unable to deal with problems when they arise. The question of what to do about those supervisors who are still not fully supportive needs to be addressed at some time.

Other issues which have required attention include: authoritarian supervisors who are reluctant to adapt to the new culture of continuous improvement and employee participation, the need for senior management to demonstrate their commitment to factory employees. Their participation in the revised manufacturing team seems to have addressed this problem. Also some of the initial project improvement projects were too large.

Impact of TQM

On management

- TQM requires considerable time and effort. Job descriptions which recognise this have been developed and include a time component for continuous improvement.
- The general manager has taken an active role in the TQM process, including attending the manufacturing team meetings.
- The factory manager now collects a considerable amount of data on various facets of the company's operations including plastic waste, machine downtime by cause and so on. These form the basis of improvement projects.

On employees

Of the company's 60 or so employees, approximately 70 per cent actively participate in teams. Administrative and sales teams are attended by relevant employees, some being members of several teams. The single production–manufacturing team is attended on average by 50 to 70 per cent of potential employees (attendance is voluntary, overtime is paid).

The positive benefits of TQM which employees have experienced include:

- working in teams
- it makes their jobs more interesting
- they are more involved with the company
- they think more about their work and jobs and have an increased awareness of things which might be improved or raised at team meetings
- it provides opportunities to speak about work-related issues
- they learn about machinery, tasks and processes
- they develop confidence as individuals in team settings
- an appreciation of the value of measurement rather than guesswork.

On suppliers

Not a great deal of attention has been focussed on quality improvement of suppliers since the main supplier is perceived to be the manufacturer of raw plastic pellets. There appears to be potential for improvement here since some of the improvement teams have discovered contaminated or faulty materials. This matter has been discussed with the relevant suppliers. One project has involved the supplier of electricity to the plant examining ways of reducing costs.

On customers

Discussions with the sales manager indicated that there was no direct measure of whether TQM had influenced sales. He felt that it was useful as a public relations device in providing quotations and general sales work. However, he did consider that in an increasingly competitive marketplace, quality is becoming an essential part of doing business. Customers, particularly those from overseas are more frequently asking about standards and quality. Tendering for any State government contracts also required quality assurance certification.

Outcomes

An obvious benefit of TQM is improved employee–management interaction and communication. This is particularly highlighted in the operation of the manufacturing process improvement team where both managerial staff and factory operators participate. During some meetings, the general manager provides a brief overview of developments in the company somewhat like team briefing. Managers who are present also hear directly from employees about shopfloor issues. Action to deal with many of these can be immediately agreed to. A further element here is that the production manager spends some time at these meetings providing knowledge about various technical aspects of the plastics manufacturing processes. Thus, the large team provides a forum for more than a single problem solving or process improvement project.

Improved job understanding, the documentation of operating procedures and the discovery of previously undetected problems have arisen from TQM and the implementation of quality assurance.

The 'leaking taps' system enables a variety of smaller issues to be dealt with on an ongoing basis. This is a very practical idea and overcomes one of the problems associated with TQM. That is, whilst one major quality improvement project is proceeding through the required PDCA process, other smaller issues can be simultaneously considered rather than having them wait until the project is completed.

Improved morale and commitment by employees is also a positive benefit. Those employees who are actively participating experience a number of positive benefits and find that they are able to input their views to initiate change and improvements.

As many organisations that have TQM in place are finding, it is difficult to isolate tangible benefits in dollar terms. Normally however, dollar savings or simplified procedures resulting from team projects can be identified. These include reduced stock holdings, improved cash flow procedures, less plastic waste and so on. Other improvement projects which have produced documented processes are difficult to quantify.

The company has maintained a register of completed projects since TQM was first introduced. A sample of some completed tasks is shown below. Many are relatively small issues which have been dealt with as a result of problems being raised at team meetings. It does not see the need to measure specific outcomes of TQM in dollar values.

Sample of TQM achievements

- purchasing done by individual departments
- separate after hours telephone line to factory canteen
- whiteboard placed in workshop to program jobs and for communication
- modified work platform in large injection
- safety check list for injection moulding
- information booklets on plastics for employees

- working capital management program
- reduction in stock holding
- shift change communication book
- reduction of plastic waste.

One 'bottom line' figure which might be used as a rough guide is that whilst sales turnover had fallen by 5 per cent in 1992, profit had remained the same. In 1993 turnover has fallen approximately 30 per cent and profit by 20 per cent. This has not deterred the push for continuous improvement.

An assessment

Within two years of the TQM initiative the program had reached an interesting phase at Nally where it had perhaps plateaued and required further impetus. The restructured manufacturing team was a relatively bold step which defied the conventional theory of group size. The team structure is working very well and has helped restore employee interest. Many factory operators have assumed greater responsibility in the improvement process.

The company is in the process of completing the requirements for standards accreditation to AS 3902. This may provide further impetus for the quality push.

Some of the difficulties which may be specific to TQM in the smaller company include:

- They have limited resources to have a full-time facilitator, coordinator or driver, ie a person who has the time to follow up procedures and keep TQM 'on the rails'. It is difficult for a single person to perform their normal duties plus manage the TQM process. This is particularly apparent in the initial years of introducing TQM. Without a facilitator, the experience was a learning one for all. Meeting frequency lapsed, participation was low, meetings became gripe sessions.

- Dealing with the 'we don't have time for this' syndrome possibly applies more to smaller organisations which are typically run on relatively lean staffing.

- There are limited resources for providing internal training and also the difficulty of attempting to introduce TQM with limited training.

The case also demonstrates some advantages which smaller enterprises can enjoy in introducing TQM including:

- Management participation on teams is highly visible. This helps to develop trust and demonstrates commitment.

- There are fewer middle managers, so often a barrier to TQM in larger companies.

- TQM is less voluntary in smaller organisations since it is feasible to involve everyone and non-volunteers are highly visible.

- Change can occur more quickly since there are fewer people to train.

- Communication and change is more rapid since managers can participate in teams.

- There is flexibility to make changes.

- Teams can deal with a number of issues rather than simply focus on one major project using the seven tools. The fact that teams are meeting and discussing issues and problems is indicative of continuous improvement where quite often problems are small and can be resolved by one person relatively quickly.

- Teamwork can be used by a relatively new company to get policies and procedures in place.

   In conclusion, this case demonstrates that TQM is not a package or program which can simply be installed as often suggested by many. It is about change and organisational development

APPENDIX: QUALITY POLICY STATEMENT

Nally (WA) Pty Ltd aims to be the acknowledged leader in the quality of goods and services it provides, and in its standards of operation.

The achievement of this aim is assured because we will:

- Implement and maintain an effective quality assurance system in accordance with the Australian and International Standards AS 3902/ISO 9002.

- Provide internal and external customers with the best value by supplying products and services which consistently meet their needs and expectations.

- Continually strive to improve consistency and efficienty in all parts of our business through a commitment to doing our work right first time.

- Encourage a spirit of pride and responsibility among all our employees for the meeting of quality objectives, as quality and reliability of the company's products and services are the concern of every individual in the organisation.

- Maintain the highest standards of safety throughout our business.

- Develop people through continual training.

- Maintain a responsible attitude towards the community and the environment which encourages open and clear communications.

Signed: ....
(General Manager)

Date: ....

This case study was prepared by Alan Brown, Edith Cowan University, Perth, Western Australia, and was first published in the *Asia Pacific Journal of Quality Management*, MCB University Press, 1993, vol 2, no 3, pp 67–76.

Case study questions

1   Evaluate the problems associated with establishing quality improvement teams at Nally (WA). How could these problems have been avoided?

2   What do you consider to be the major challenges facing small companies in implementing TQM?

3   Is the TQM implementation approach any different for small companies compared with large companies?

C5

Total Quality Culture at Colonial Mutual Fiji

We decided early in our quality improvement efforts to label our new management approach TQC. Total Quality Culture (TQC) best represented our vision of what we needed to change and improve, to achieve continuous improvements and quality products and services to our customers. We do business in a multi-ethnic society, with Fijians, Indians and other minority races on both sides of our employee–customer equation, we want to create a company that will complement this diversity.

The beginning

As an organisation, Colonial Mutual Fiji, and indeed Colonial Mutual as a Group have always sought to assure quality service for our customers.

We went about it in various ways such as through training, close monitoring of control charts, showing turn-around time of queries and correspondence, occasionally raising awareness of the need for quality service through staff magazines, meetings and seminars.

A procedure, once established became sacrosanct. When a complaint arose the person or the department involved was reprimanded. Training and more training was the solution prescribed to most problems.

People grew up in the organisation with established rules and processes which have never changed. These processes became revered and required the intervention of no less than the CEO to get changed.

Customers who required service at their more convenient hours of say 8.30 am to 9.30 am or 3.30 pm to 4.30 pm were denied such services because the organisation counter-service rule said that counter services opened between the hours of 9.30 am to 3.30 pm each day.

Customers who urgently needed loan cheques in the afternoon had to wait until the next day because the rule stipulated that cheque machines stopped processing at 3 pm each day.

Rules and procedures ruled the day. Service and customers had to fit within these parameters. We made commitments to provide Quality Service and expected our people to achieve this. Performance appraisal and remuneration were based on how well our people understood and worked within the boundaries of established rules and procedures.

As the CEO of Colonial Mutual Fiji, I began to show interest in Quality Management. I realised its potential for Colonial Mutual Fiji since I looked at it as a structured and disciplined way of ensuring quality service to our customers. Colonial Mutual Life Australia (CMLA) had by this time implemented TQM. I was therefore able to discuss with CMLA people various aspects of TQM and to observe first hand TQM at work.

Getting started

I knew I had to use someone to assist me in implementing TQM in Colonial Mutual Fiji. It was therefore a question of deciding who to get and from where.

I was quite certain of what to get. My objective was to find a consultant who understood Fiji and the cultural mix in existence there. The consultant was to help design a plan which took into account the cultures of the people and the culture of the organisation. Two important messages are found in that last sentence.

Firstly the consultant had to work with our people in designing TQM for the organisation so that they knew that they were part of the proposed change. Too often TQM 'packages' are 'picked off the shelf' and are expected to work in another organisation without taking into account the local culture, values and attitudes.

Secondly what works in Australia may not necessarily work in Fiji because of cultural differences and furthermore what works in one company in Australia may not work for another in the same country.

I had heard that Jerry Glover, Professor of TQM at Hawaii Pacific University, had been working with the Castaway Resort in Fiji. I met him briefly in Nadi and arranged to meet him in Hawaii where at that time he was assisting Hilton Hotels to implement a quality improvement system. At a further meeting in Sydney three months later we discussed what had to happen to make our quality efforts successful.

First of all, I had to be the leader of all quality efforts. No delegation or passing off responsibility to subordinates. I had no problem understanding that quality required my direct involvement since it is an operational and strategic means of running Colonial Mutual Insurance.

We also discussed how we must design a quality system to fit our location and circumstance of operation. Fijians have long been accustomed to working in groups, so the team approach appeared to fit. We also considered the most appropriate implementation plan. As it unfolded, it was a very ambitious and thorough approach. The key ingredient was the decision to begin with our senior management people and develop them as our internal expertise.

Implementation

We planned for a total organisation commitment to Total Quality Training that lasted for two weeks. With my involvement, Jerry Glover conducted a five-day working session with all managers. The goals of the session were:

- to provide an overview of TQM
- to develop the managers knowledge of Total Quality
- to develop a working understanding of TQM methods via participative process engineering teams
- to develop realisation of what Total Quality can actually achieve for our organisation
- discussion of organisational culture shifts
- establishing agreement on what our efforts would lead to in the form of measurable goals.

The following week, the rest of the employees were trained on Total Quality. The presentation of TQC to employees was actually done by their Senior Managers. I acknowledged that unless the concept was widely accepted across the organisation, I would not be able to arrive at the desirable culture shifts for TQC to work.

TQC design and implementation plan

A Steering Committee comprising of all managers was formed and members of this committee acted as trainers and facilitators in our continuing effort to involve all staff in TQC.

Our design made use of the existing hierarchical and functional structure. We created a team concept and team identification in Colonial Mutual with a three-tiered team system.

This team system accomplishes our goals of 'fixing' both macro and micro-processes. However, we have encountered very little resistance to change since we did not design the team system to be ad hoc or multi-unit at the employee level (cross-functional). That is, regardless of the three levels, team members always have the functional authority (and vested interest) to fix the process. We also felt it important that the teams select process problems which they perceived to be important. Each team was initially assigned a facilitator appointed by the Steering Committee.

Kaizen is an important influence on our team design. By structuring our employee teams in a functional manner, the members concentrated on the micro-processes (cockroaches) which tend to interfere with their daily work and service to the customer. On the occasions when the team identifies a macro-process (elephants) to be fixed, the problem is referred to the appropriate middle management team or if it is company-wide, to the Steering Committee. With our team structure, we can focus on 'Kaizen-type' problems, while allowing the managers of the organisation to continue to work on leading our efforts in a manner and functional structure to which they have been accustomed. A team approach is typical of Fijian rural administration structure.

We have also realised that 'broken' processes are symptoms of our real problem – years of operating within a traditional Western organisational culture. Our goal is to create an adaptive, proactive and 'continually learning' organisation capable of meeting the needs of our customers in our unique society. We envision our typical workplace in the immediate future as an environment operating in a horizontal manner. Employees will be formed in self-directed work teams whose daily activities are focused on keeping their work processes in control. On the wall, we expect to have process control charts which will provide us with real time measures of our performance and customer satisfaction levels. The employee work teams, their supervisors, and our managers will be recognised and rewarded in relation to how well our processes are managed.

The Steering Committee meets on a monthly schedule for two to three hours during which they also receive process improvement presented from Quality Teams. The other teams meet each week. Teams that have solved a process problem are presented with a certificate. This certificate acknowledges that the team has been properly supported in its efforts and that the team has properly identified, analysed, and documented the process re-engineering effort.

We have been able to integrate our existing organisational culture with quality management, and by *cakacaka vata* (working together) we have developed a more appropriate sense of place for our work environment.

TQC in progress

In the initial phase of the implementation we progressed at a 'controlled pace'. We created many changes in our organisation, all directed at our goal of continuous improvement. Some of these accomplishments include:

• Our senior managers have taken credit for the success of our TQC effort (in fact, they were the ones who labelled our efforts Total Quality Culture in recognition of our real goals).

• Twenty-five quality teams have been established and these have solved process problems (we established functional teams in each operational unit while forming cross-functional middle management teams).

• Fourteen teams have resolved/improved processes using the structured problem solving steps.

• We have implemented certain organisation structure changes to be more responsive to customer needs.

In less than one year we created the foundation for our Total Quality Culture. This foundation has resulted from a transformation of our original organisational culture, structured according to the traditional Western paradigm, to our new paradigm based on our quality improvement principles. That is, many of our new organisational structures (TQC) resemble the way we have lived and worked in Fijian society for generations.

We have accomplished this progress primarily because, as senior managers, we have realised both the necessity for organisational culture change as well as the importance of our roles as leaders of change. Quality improvement results only when sustainable structural changes are made in the way an organisation does things on a daily basis. We have questioned all of our previously held assumptions about organisation and management. No practice has been considered 'untouchable'. Our information technology system has been restructured to be process focused, we have modified our performance evaluation approach to be more team-oriented, and we have actually changed our decision-making structure at all levels. These changes are not easy and require a significant investment of time on senior management's part. We had to first know what our current practice was, then design a new structure to fit our TQC.

The Steering Committee has been the primary source of organisational changes.The Steering Committee created an action plan to make certain changes. Key features of this were:

- mapping of Colonial Mutual's macro-processes

- measurement systems for the macro-processes

- identification by employee teams of micro-processes

- measurement systems for micro-processes

- development of an integrated process/job description system

- a new performance appraisal system linking individual and team evaluation and recognition.

Challenges we encountered

The adoption of Total Quality meant a significant organisational change. Cultural change, in particular, was difficult but we saw and accepted the need to commit our resources to facilitate the desired changes. Challenges we encountered were:

- Change to completely new thinking and habits takes time. It will not happen immediately despite early concept acceptance and understanding.

- TQC effort and activity if not persistent and rhythmical allow old bad habits to retard it.

- Pace of progress can be slow due to consensus based team decisions with team speed being determined by the slowest member ie. convoy fashion.

- If not high on operational agenda at any level it suffers lethargy and can be counter-productive.

- Initially, not all staff were involved in problem solving teams and those left out tend to isolate themselves from the concept.

- That there is no objective benchmark to measure progress against.

- Staff initially tend to consider TQC as a 'project perception' rather than being 'part of daily activity'.

- Difficulty initially in allotting time allowance to attend TQC meetings and related activities without interfering with the usual workflow requirements.

- As we got into Quality Management it seemed that, now and again, we were not going forward in terms of work quality and output. During the early stage of the implementation we occasionally had doubts as to the practicability of Quality

Management. However, we had to keep reminding ourselves that Quality
Management does not produce instant success and that we had to persevere and take
the long term view.

As the first organisation in Fiji to embrace Quality Management, Colonial Mutual
Fiji was alone with no-one to turn to for encouragement and support. The implementation of quality management meant that we were taking the organisation into the future,
setting a new direction and a new way of doing things.

During the implementation period as we considered the enormous but necessary
cultural shift of the organisation there were moments of doubt whether we were taking
the correct action by steering the organisation into unchartered waters which might
create confusion, cause low morale and bring ultimate disaster.

Fortunately management and staff believe that we have taken the right course and
they are certainly proud of it. Their enthusiasm and support sustained and encouraged
me greatly.

The diffusion of TQC

Our goal was to have our design for TQC be the most appropriate for Colonial Mutual
in Fiji. The original design was modified in some ways as we progressed through our
early implementation efforts. For example, there were design considerations that we
addressed after beginning our implementation, such as how to modify our performance
appraisal system to be more compatible with team motivation and recognition.

When we developed our implementation plan we were aware that we had to modify
our 'Western organisational culture' to permit TQC to begin to have positive results. In
fact, we saw TQC as an innovation of a social and organisational nature. As such, we
were as concerned with our concepts and methods for creating organisational change as
we were with the statistical process methods and other more commonplace quality
tools. Innovation diffusion or how a new concept or idea is spread and adopted in an
organisation provided the direction for TQC.

Rogers (1983) in his classic work, 'Diffusion of Innovations', explains the need for
us to understand new practices such as quality management as innovations to be
diffused into an organisation's social system and its culture. Particularly important to
successful acceptance of an innovation are its attributes. Research on innovation attributes has given us some degree of confidence in predicting future successes. We were
guided in our efforts to have the members of our organisation adopt TQC by an implementation plan which was structured according to five implement innovative attributes
cited by Rogers as the most important:
• relative advantage
• compatibility
• observability
• complexity
• trialability.

Relative advantage

We discussed how TQC would fit our Fijian cultures during the Senior Management Training Program. Our delight was that many of the organisational requirements for quality improvement are familiar to us. Group affiliation, consensual decision-making, and reciprocal economic exchanges which involve trust are central features of our tradition. We were interested to see the close similarities of our traditional ways with the new globally popular quality management approach.

This conceptual link was important to our acceptance of the new quality approach. We perceived a relative advantage for the new innovation of quality management in several other ways. Colonial Mutual is a service business and our customers are our priority. However, we realised early in the implementation that we had developed some bureaucratic processes which were not serving our customers as well as we should. Quality improvement teams gave us a new means for fixing those processes and providing better service to customers.

Another advantage of the new quality effort is that our design for TQC did not create insecurity or uncertainty that so often accompanies change. For example, our new team structure created a more effective hierarchy and process engineering capability without alienating our existing chain of command. Senior managers worked as the steering committee, middle managers dealt with cross-functional processes. We were able to adopt the new team structure within a framework that did not modify our daily work expectations.

Compatibility

Our TQC innovation was familiar to us, and as such, created a psychological support throughout the change process. Also contributing to TQC's compatibility was its compatible nature to what we were already doing at Colonial Mutual in focusing on customer service. Our methods changed because of TQC, but our priorities of customer service did not.

Complexity

We also attempted to keep our TQC innovation free of unnecessary complexity. Our design was explained to all levels of management and staff so they could understand their roles and contribution to TQC. Also, our horizontal management approach is designed to eliminate any bureaucratic complexity in responding to market needs.

Trialability

TQC was easy to try for all Colonial Mutual staff. Our initial training required everyone's attendance. More importantly, our Steering Committee and facilitators have 'orchestrated' early successes in our teams to ensure that everyone has achieved success with TQC. It is important for potential adopters to try TQC with positive initial experience. Our facilitators have supported team efforts as have our operational managers.

Observability

TQC's early successes have been observed by all our people at Colonial Mutual in the form of processes improved by our quality teams, certificate presentations to the Steering Committee, and *kava* sessions at the end of the day to discuss TQC merits with staff. Perhaps the most important signal given to our people at Colonial Mutual was the full involvement of our senior managers and myself at all TQC events and activities. We have sent a message to everyone at Colonial Mutual that TQC is not only our vision, but also our operational reality.

Acknowledgments

Professor Jerry Glover, Professor Tusi Avegalio, Lisa Mosher, Laurie Foster

This case study was prepared by Tom Vuetilovoni, Colonial Mutual Fiji.

References

Rogers, E.M., Diffusion of Innovations, 3rd Edn Collier MacMillan, London, 1983.

Case study questions

1   Evaluate the approach adopted by Colonial Mutual Fiji in relation to their TQC initiative.

2   Discuss the role of a steering committee in a TQM program.

3   What advantages did Colonial Mutual Fiji have in changing to a TQC organisation?

C6

Achieving a Total Customer Service Culture: The Hong Kong Mass Transit Railway Corporation Experience

Introduction

Hong Kong is a small area of land on the southern coast of China of only 1070 sq km. It has developed from a fishing village in the 1840s to its present day status as an international business centre where East meets West. It is one of the world's most densely populated cities with a population of more than six million. It has an annual gross domestic product per capita which is comparable to many western countries.

Only a small proportion of the population own private cars. Consequently the transport scene in Hong Kong is dominated by the public transport system.

The Mass Transit Railway (MTR) was built into this densely populated area and started operating in 1979. It has become an integral part of Hong Kong's daily life. The total route length of the system is 43.2 km covering major commercial, social and residential areas of Hong Kong. The system comprises three separate lines with cross-platform integration: Kwun Tong, Tsuen Wan and Island Lines. There are 38 stations, 30 of which are underground. Every weekday the MTR carries over 2.3 million passengers.

The Mass Transit Railway Corporation (MTRC) employs 6000 staff, 4000 of them engaged in the operating/operations engineering departments. The rest of the staff work in finance, personnel and administration, marketing and planning, estate management or the project departments.

Shaping the culture

Experience indicates that companies which are truly excellent abide by a small number of fundamental values which are fully understood and endorsed by employees at all levels. In view of this, in 1985, the Corporation embarked on the process of shaping the values and developing the strategies and objectives of the MTRC in a bid to effect changes in the systems and practices, and in the behaviour of its employees.

Setting the values

After a process of considerable consultations on the established beliefs and practices against the changing business needs, three core values were defined and launched as the guide for shaping the corporate culture and behaviour. These are:

1   *Customer Service.* The Corporation aims to be the most customer-orientated railway in the world, and to provide by far the most reliable service and to strive for continuous improvements to meet or exceed the needs of fare-paying passengers.
2   *Respect for the Individual.* In all dealings with and between its employees, the Corporation will ensure that high standards of fairness, equality, objectivity and empathy will be observed and that the needs, opinions and aspirations of all parties are mutually respected.
3   *On Time and Within Budget.* The Corporation has responsibilities to shareholders, lenders, customers and staff to operate professionally and in a financially prudent way to achieve the most effective and timely use of resources.

The MTRC's three core values reinforce one another. The first, customer service, puts emphasis on the importance of providing excellent service to its customers. Over two million passengers take the MTR every day. It is the Corporation's responsibility to provide a superior service both now and far into the future so that customers can depend upon the reliability of every aspect of the service provided. Frontline staff are trained and encouraged to provide their best service to meet customer's needs. All staff are encouraged to take this same attitude towards their internal customers.

The second value, respect for the individual, is the essential counterpart of the first value. An organisation can achieve results only when its employees are committed, happy and feel proud of their work. The Corporation believes that only when it treats its employees with respect and empathy can it expect loyalty, commitment and a customer focused attitude in return.

The final core value, on time and within budget, conveys to all staff the importance of meeting targets and achieving results with high efficiency and cost-effectiveness. Just as with many reputable railways around the world, financial success is a vital ingredient to maintaining the long-term safety and service standards. This is therefore a critical value that supports the customer service strategy of providing a reliable and efficient transport system for the community of Hong Kong.

Diffusing the vision and values

Once the core values were defined, an intensive communication programme was designed to promote awareness throughout the Corporation. The message explaining their importance and meaning was delivered in a dramatic and creative manner through videos, songs, handbooks, various core value competitions, company dinners that included husbands and wives, and many other family events.

Aligning behaviour

With sufficient awareness, a roll-down training strategy was adopted to ensure that every member of staff is equipped with the right attitude, knowledge and skills to support the Corporation's commitment to its customers.

The core values were communicated to staff at all levels through managerial training programs, internal customer service training, supervisory training programs and intensive skill training workshops. A specific program called 'Vision Day' was designed for frontline and junior support staff to explain the concept and importance of customer service, in such a way as to align their commitment to delivering service excellence. Focus Days were organised for managers to provide a forum for discussion on internal customer service.

Setting targets and standards

The training initiatives were supported by a comprehensive set of internal and external customer service targets and standards which were established to ensure that employees had a clear understanding of the customers needs and management expectations of performance.

By referring to the international quality system standard, ISO 9000, a Quality Management System and a Safety Management System were established to ensure that the maintenance standard and services delivered meet the international quality and safety requirements as well as the customer service standards and targets.

The Corporation has undertaken various customer service projects in striving for continuous improvement based on customer feedback and expectation. These projects include the acquisition of additional rolling stock and the enhancement of the signalling and control system in order to improve train service; the development of high-coercivity tickets to improve ticket reliability; upgrading of train announcements by pre-recorded digitised voice announcements to improve the communication system with the customer, and enhancement of maintenance to improve escalator availability.

Recruiting and training the right people

Customer service is incorporated as a focus in the recruitment and selection process. Since 1992, the following lead-in paragraph became mandatory in all MTRC recruitment advertisements.

> The Hong Kong Mass Transit Railway Corporation is a customer-orientated organisation dedicated to reliability. For those who share our commitment, we offer good training and career opportunities.

Specifically designed selection tests were introduced in 1994 to select customer-orientated Train Operators and Station Staff.

All newly recruited frontline staff have to go through an intensive basic training programme that covers technical subjects as well as customer service skills.

The initial orientation and training are further enhanced by a whole range of internal training courses covering a wide range of managerial, supervisory and professional skills. The Corporation sponsors employees for relevant external continuing education. This is to ensure that staff are equipped with the necessary knowledge and skills for their jobs and to support the MTRC philosophy of continuous improvement through learning.

Maintaining good performance

Staff are motivated by the recognition and reinforcement of positive performance through a number of motivational schemes. These include the Chairman's Award, a

Long Service Award, a Staff Suggestion Scheme and an Encouragement Award. Since the introduction of the Encouragement Award in 1991, about one-quarter of the awards given relate to customer service.

The Corporation has developed a sophisticated performance management system called 'Managing Performance For Success'. This promotes a joint accountability between supervisor and subordinate for the successful accomplishment of work-related targets and the development of career plans. To take care of the career development of its employees, Development Centres are used to enable staff to identify their potential and their career aspirations.

Communicating internally

The Corporation proactively promotes two-way communication between management and staff by conducting internal surveys. A company-wide Staff Attitude Survey is conducted every three years to measure the general climate and morale of the company.

There is also an Internal Customer Service Survey which collects staff feedback on the service quality provided by various internal departments and the areas that need improvement. This survey provides data for discussions amongst staff in the same department or from different departments to further improve the organisation systems, processes, human resource activities and provision of internal service.

Frank and constructive dialogue between management and staff is further enhanced through consultation at the Joint Consultative Council and the Staff Consultative Council which was formed amongst staff at different levels.

The Corporation's culture and values are reinforced by various means of communication including a Work Improvement Team Scheme, a video newsletter published every three months, regular publications such as *The Express*, and a yearly *Performance Review*.

Effective two-way communication within the MTRC is particularly important as management believes in mutual respect and have a long-term vision of continuous improvement.

Communicating with passengers

Customer-orientation is considered to be essential for success. In order to determine customer needs and attitudes, the Corporation has, over the past 13 years, conducted over 180 research projects which have involved over 700,000 respondents. The results are used to identify trends, determine needs and set customer service targets.

To obtain views directly from passengers, passenger liaison meetings in the form of Coffee Evenings where passengers can talk to MTRC management staff face to face are organised on a monthly basis. Other support services for customers include the MTR hotline and provision of passenger suggestion boxes at stations.

It is the Corporation's commitment to provide correct, updated information to passengers. Publications are regularly compiled to ensure availability of MTR information. This is demonstrated by the four monthly publication of the Customer Service Report, an honest and open appraisal of the Railway performance against three challenging service reliability targets, for Trains, Tickets and Escalators. Such a commitment to customer service performance standards has been the first by a public transport operator in Hong Kong.

The image of customer orientation has been projected through advertising campaigns and promotions. The Corporation proactively communicates with political, commercial and social communities. This is done through press conferences, seminars and meetings with District Board members.

Recognitions and achievements

Despite strong competition and without expanding its system, the MTRC has achieved a competitive advantage over Hong Kong's other transport providers. This is reflected in a steady increase in market share over the years. In 1992, the MTRC market share stood at 27.1 per cent of the total public transport business.

Financially, despite an outstanding debt, the Corporation had a net profit of HK$400 million after interest and depreciation in 1992, and anticipates retiring its debt by the year 2001. It is one of the few profitable railways in the world, and is the only credit rated public transport company in Hong Kong with an investment grade credit rating which enables the Corporation to borrow at competitive cost margins.

With respect to employee relations the Corporation enjoys a relatively low turnover rate of around 6.5 per cent compared with the Hong Kong average of 24.5 per cent (Hong Kong Institute of Personnel Management), and a comparatively low absence rate of 3.6 per person per annum (Employers' Federation Survey 1992). A recent Staff Attitude Survey revealed that three quarters of all employees considered the Corporation to be a good employer.

The MTRC's dedication to achieving a total customer service culture is recognized by various awards it has acquired. For two consecutive years (1991 and 1992), the MTRC won the Hong Kong Management Association's 'Award For Excellence in Training' and the Chairman, Mr Hamish Mathers, was named as the 'Communicator of the Year 1992' by the International Association of Business Communicators in recognition of his and the Corporation's achievement in the effective and open communication both with its staff and with the general public. The MTRC won further recognition when the integrated marketing communication programme 'MTR Customer Service Campaign 1992' won the 'Hong Kong Management Association/Television Broadcast Ltd Marketing For Excellence' Award.

This case study was prepared by Sara Tang and Patrick Maula, Mass Transit Railway Corporation, Hong Kong.

Case study questions

1   Evaluate the case study in terms of the vision and values adopted and their diffusion within the organisation.

2   What role did communication play in Mass Transit Railway Corporation's quality initiative?

C7

Company Culture and Total Quality Management: A Case Study

Introduction

The global market in the 1990s is ever-changing. The advancement in technologies and product sophistication provide a breeding ground for a vast number of ambitious entrepreneurs.

Some companies failed to withstand the technology revolutions, the changes of the market needs, and the emergence of new competitors. Some were trapped in the pitfall of their own temporary success and advanced no further. Few could catch the rhythm of the era and geared up for continuous business growth.

Why did some companies flourish for a while and fade without trace, while some march along with the era with ever-lasting strength? The key of success lies profoundly in 'company culture'.

With some innovations, entrepreneurs may find it not too difficult to establish themselves in the market. However, in the time of market downtrend, change of customers' taste, or personnel changes within the companies, their integrity will be put to the test. Some of the vulnerable ones will be out of the game, but those who can stand will view this as a favourable natural selection, and turn the crisis into opportunity.

Just as each individual has his own personality each company has its own culture. The company culture is the integral behaviour of the various factors within the organisation, including both external and internal factors. Every employee brings his own knowledge, experience and value into the company, and incorporates all these into the operation of the organisation.

The culture of a company may contain both good and bad ingredients. These ingredients evolve in the course of the development of the company and became an integral part of the company.

Every company has its own value system which may be developed unconsciously or may be purposely designed. In most cases, the original culture of a company is significantly influenced by the founding members. Through interactions among old and new employees, and the continual cooperation between the employees and the organisation, the employees are conditioned to adopt the value system of the company. On the other hand, company culture is very much affected by the behaviour and performance of the employees.

Company culture is not only reflected in the group behaviour of the employees, but also in other norms, such as work area layout, messages on the notice board, parking space arrangement, the way of answering the telephone, tidiness of the cafeteria. These activities, though trivial, usually reveal the value system within the company.

A well-established company culture can intake new ideas to fuse into its system for further improvement and continual renewal. However, changing the culture is easier

292

said than done. The company culture may have been there for a long time. The beliefs of the founding members and the top management team are translated into rules and systems, norms, style of managing and so on. These are passed on to people who joined the company. Part of the culture may well complement the concepts of Total Quality Management (TQM) or perhaps part may contradict.

The latter can be reflected in a lot of symptoms including poor layout, bad working environment, or hostile human relationships. This can be further reflected in the behaviour of the management team, which is erratic and geared to achieving short-term results, spending efforts on fire fighting instead of preventive measures, etc. In order to make TQM successful, these attitudes and behaviours have to be uprooted.

The success of TQM relies on the participation of all employees. If the TQM activities are incompatible with the company culture, the implementation will be difficult. Hence the management team has to change the company culture deliberately to enable employees to contribute to the TQM program, and provide an environment where involvement in problem-solving and decision making is the norm. Any group norm opposing this principle should be changed. The most important factor in this effort is leadership. Without total commitment by the leader of the company, it is impossible to enlist the commitment of all employees to TQM.

There are common cultural characteristics of most excellent companies. Peters and Waterman (1984) identified that 'hands-on, value driven' was the characteristic which mostly reflected the importance given to company culture. Some sample beliefs are:

- emphasis in the importance of enjoying one's work
- a belief in being the best
- people are encouraged to innovate and take risks, without fear that they will be punished if they fail
- attending to details
- respect people as individuals
- a belief in superior quality and service
- improvement of flow of communication
- importance of 'hands-on' management; not just planners and administrators
- importance of a recognised organisational philosophy developed and supported by top management team.

In order to re-build the company culture, we have to first set up a common value – a stable and long-term mission. This mission should be shared by the employees, whose participation and contribution are emphasised. The following case study illustrates these points.

Managing change – A case study

Computer Products Asia–Pacific Ltd (trading as Power Conversion Asia–Pacific, PCAP) was established in Hong Kong in 1981 as a manufacturer of power supplies for the computer, telecommunication and equipment industries. When we progressed towards the end of 1990s, the framework of our management system had been well established. We were proud of our team of hard-working and dedicated employees. However, they were constrained to the framed system and hesitated to suggest changes.

In order that the company could further upgrade itself and best meet the needs of the ever-changing electronics market, we decided that we should rally our employees around a common vision. Hence we promoted the concept of '3C':

- charge
- change
- challenge.

In order to achieve top-quality products and superior performance, PCAP encourages all its staff to 'charge' themselves through attending in-house and external training seminars and workshops designed to equip them with further knowledge and skills. Employees are expected to accept 'challenge' with responsibility and courage, and to have the ability to manage 'change' in order to improve.

The '3C' was not just a slogan to make it part of our company culture, we encouraged our employees to visit other factories to see how they were managed. We invited experts from outside to share with us their professional experience. We employed a professional training manager to design and organise an integrated employee training and development program. These series of actions realised our commitment to 'charge'.

We encouraged our employees to air their suggestions for improvement and innovation. In the past, our design centres were in the USA and Europe. The plant in Hong Kong was a manufacturing centre. We followed the instruction provided by the design centres. Ideas for product improvement only incurred when there were serious manufacturing difficulties. Encouraged by the concept of 'challenge', our employees started to challenge themselves with a higher standard of product quality. Not only did they contribute on improvement of production methods, but also initiated product design changes. We aimed to develop products with high manufacturability, high production yield, lower customer return rate, shorter cycle time and lower production cost. After only eight years of operation the Hong Kong plant was upgraded to be a business division and Asia-Pacific headquarters. Our engineering team formed a 'concurrent engineering' circle with their counterparts in USA and Europe to develop products for the global power conversion market. As a result, production yield and customer return rate were improved, development cycle was reduced. Furthermore, our employees were more confident. They were ready to face any challenge and to work out the solution.

The third concept, 'change,' was also shared among our employees. We accepted continuous changes and improvement as the way of life. The award of ISO-9001 Certificate highlighted the importance of 'change' in the development of our new and complete quality management system.

In the first six weeks of our ISO-9001 program, we worked together with our consultant to perform a detailed analysis of our existing quality system. From this analysis, we constructed an improvement plan with detailed schedule of tasks and manpower allocation. The first step of this project was to identify the scope of application for ISO-9001 which included the types of product or services, and affected departments, sections and locations.

During this stage, we started to promote the spirit of ISO-9001 among the top and middle management, to help them understand the role of every employee and every department in the ISO-9001 system. We set up a task force with representatives from all departments. The chairman of the taskforce was the quality system manager (acting as quality manager) who directly reported to the managing director. No additional staff was hired for the ISO-9001 program throughout the whole project.

In the second stage, we focussed on rewriting the Company Operation Procedures (COP) and Working Instructions (WI). Although there were many established procedures which did not need dramatic changes, it was hard work to rewrite them in order to align these with the ISO-9001 requirements. After the completion of the first draft of COP and WI , they were discussed, checked and amended several times to make sure they were practical. Moreover, the quality system manager and the consultant had compared the documents carefully with the ISO-9001 standard to ensure that all the specified requirements were met.

Since our colleagues had to take care of their day-to-day responsibilities, they spent extra time and worked long hours to prepare the procedures. Nevertheless we were proud of the dedicated documents which were delivered on time. Members of the task force not only had to keep monitoring their departments' operations to meet the new requirements, but also had to perform internal audits to other departments. They provided recommendations and served a monitoring role for the system. Through these exercises, the cooperation among colleagues in different departments was improved because they understood the problems other people faced.

In less than two years of starting our '3C' initiative we had completely edited our quality manual. PCAP's quality policy statement was also finalised. The '3C' principle was incorporated in the statement, with the goal to 'do it right the first time, and every time'. Following the release of the quality manual, 'Company Operation Procedures and Working Instructions', we entered the third stage – implementation of the new quality system. The process involved a total mobilisation of people. We organised the employees into many units, provided training courses to them, made them familiar with the new system, and ensured that what we were doing conformed with what we put down. At the same time, we amended documents to eliminate the contradiction among the procedures. Relevant data and records were collected in this stage to prove the system operated effectively.

In order to examine the execution of the new quality system, internal audits were also carried out. A team of trained internal quality system auditors was appointed to find out any discrepancies between the documented system and the actual practices. At the same time, our consultant also completed a quality system review for us.

We also invited the Hong Kong Quality Assurance Agency (HKQAA) to conduct the audit. The audit revealed some deficiencies in our document control. Corrective actions were taken immediately. Within four months time, not only did we successfully rectify all the defects but we also completed another detailed and comprehensive internal system audit. Finally, we were full of confidence to invite the HKQAA to come again for the re-audit.

HKQAA conducted a three-day in-depth assessment on our quality system. All employees were pleased when the HKQAA audit team announced on the third day that we had satisfied the requirements for ISO-9001 certification, and became the first Hong Kong company to be awarded with ISO-9001 certification.

The whole process for preparation and certification of ISO-9001 took only 12 months. It was mainly contributed by a strong commitment of the top management team and the participation of all employees. During the 12 months period, we had successfully restructured and enhanced our quality system. Our '3C' principle was a strong driving force.

Our employees were encouraged to 'charge' themselves through learning and training, to accept the ISO-9001 'challenge' and to build up the ability to manage 'change'. Besides, the demands of the ISO-9001 quality management system is in fact are very similar to our beliefs. Hence, we were able to obtain the certification smoothly and in a comparatively short period of time.

Company culture cannot be established overnight, nor can it be changed immediately. The leader must become not only an advocate for the culture but also a role model.

References

Peters T.J. & Waterman, RH, *In Search of Excellence: Lessons from America's best run companies*, Harper & Row, Sydney, 1984.

This case study was prepared by Lo Wai-Kwok and Tong Wai-Kwok, Computer Products Asia–Pacific Ltd, Hong Kong

Case study questions

1   What were the key features of the quality initiative at Computer Products Asia–Pacific Ltd? Were these sufficient in creating a TQM culture?

2   Based on the information given in the case study, discuss the impact of ISO 9001 certification on company culture.

C8

Quality in Information Technology: Incorporating the Market Driven Quality Concepts

Introduction

Statistics released by the Association of the Computer Industry Malaysia (Pikom) revealed that the size of the information technology (IT) industry in Malaysia in 1992 totals RM2.03 billion. This encompasses equipment for processing, communicating and storing of information; the programs and systems that enable these equipment to be applied to specific business processes; and the services related to facilitating the acquisition, development, operation, and maintenance of these applications.

The long-term potential for the industry has been rated most attractive given the contention that for Malaysia to make the required quantum leap in order to compete globally and achieve the goal of being an information rich society as envisioned in our Vision 2020, we will need to increase IT investments per capital from about 0.6 per cent in 1990 to match that of the other more developed countries; 1.0 per cent for Taiwan, 1.25 per cent for Singapore, 2.0 per cent for Japan and 3 per cent for the United States.

The double digit growth recorded by the industry over the last few years has attracted a large number of new entrants into the various segments of this industry. A recent survey by IBM Malaysia showed that there were over 1500 companies involved in the industry in the country.

As a result, competition for the client's IT budget has been most intense. Market players have responded to tap these opportunities by bringing in new and state-of-the-art technologies, innovating value-added services offerings, investing in strategic alliances with other key players and customers, and out-bidding prices in attempts to maintain or gain market share.

Quality has been a key thrust of the industry in Malaysia. Major industry players understand that quality is critical to ensure that the industry can mature to its potential. In fact, many are positioning quality as a competitive weapon to build longer term partnerships with customers. As a leader in the local IT market, IBM Malaysia has been in the forefront of these efforts.

IBM's quality initiative is termed MDQ or Market Driven Quality. The basic MDQ premise is that if one can be the best at satisfying customers in the chosen markets, then everything else important will follow. Its truism lies in the underlying commitment that since the customer is the final arbiter of the business, quality is nothing more than satisfying the customer efficiently.

This case study attempts to document IBM Malaysia's experiences and learning in implementing this quality initiative. The last five years had indeed been challenging and rewarding for us here in Malaysia. We undertook the creative destruction of our past successes; pro-actively changing everything to re-invent ourselves for our customers.

IBM Malaysia quality drive

IBM Malaysia has the distinction of leading the rest of the IBM Corporation. IBM Malaysia became the first marketing and services organisation in IBM worldwide to exceed the silver and then the gold levels of the IBM Corporation quality assessment in 1992 and 1993 respectively. The assessment, modelled after the US Malcolm Baldrige National Quality Award, benchmark the progress made in the transformation of IBM Malaysia into a market driven quality (MDQ) organisation. This leadership was also recognised by the Malaysian government in 1992 when IBM Malaysia was awarded the National Award for Quality Management.

We are fully convinced that our success this time around is without doubt attributable to the total quality view we adopted, as opposed to piecemeal approaches of prior efforts. The implementation framework is built around a transformation model that is targeted at effecting a complete behavioural shift in line with the desired market driven quality paradigms. This is because we believe that behavior is the final manifestation of our interaction with the customer (often called moments of truths).

We realise this total perspective by incorporating a simple behavior model (see Figure 1), based on the works of Brown (1982). The objective is to orchestrate the desired behaviour by realigning three key components – the mind-set, the determining elements and the reinforcement factors.

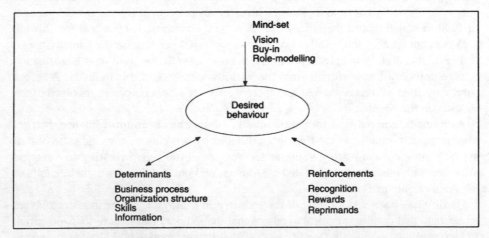

Figure 1   *Behaviour model*

Changing the mind-set

The mind-set is perhaps the biggest challenge in effecting this change. Our experience pointed to almost a third of the last five years were spent on getting the required buy-in, even before we could delve into actually synchronising the determinants and reinforcement factors.

Getting our people to accept the need to change is a major undertaking indeed given that IBM Malaysia was and is a successful business concern. While we anticipated that we may fall into the trap of initiating this change too late, if we were to wait for later when the growth of our business starts to level off, we discovered that our past successes can in many ways be a major hindrance to our transformation.

To set the tone, we announced right from the very beginning that anybody in the company who could not relate the things they were doing to delivering values to the customer would immediately cease doing those activities. This had an immediate impact in re-orienting the perspective of everyone to the customer. In the same way, we discontinued the appointment of a dedicated quality manager. Instead, we declared every employee to be a quality manager in order to promote ownership.

Surveys we conducted to tell us the acceptance of MDQ by our employees also pointed to a disturbing phenomena at the early stage which could very well have frustrated the change effort had it not been rightfully addressed. These surveys measured the view of the employees with regard to their commitment and that of their peers and managers (Table 1).

Table 1   MDQ acceptance by employees

Year	Are you committed to MDQ? %	Do you think your manager is committed to MDQ? %	Do you think your colleague is committed to MDQ? %
1	83	48	21
2	96	93	84
3	98	97	89
4	99	94	92

It was obvious that most employees rated themselves to be highly committed right from day one but interestingly not their managers or colleagues. This 'I'm OK but you're not' perception caused many employees to stay with the status quo because they do not see the need to change the way they operate since they perceived the problem to always be with another person and/or function. We noticed that nothing in fact changed over this time despite the commitment and good intentions. It took us almost another full year to re-position this perception, and thus get their trust and confidence, before we got the real buy-in.

Our model contends that the desired behavior cannot be sustained by the mind set alone. Key behaviour determinants are required to mould the necessary predicability and repeatability qualities to enable us to satisfy the customer each and every time. This will then have to be reinforced by relevant factors that recognise and reward the right behaviors and/or reprimand the wrong behaviors.

We see the key behavior determinant to be the business process, rather than the organisation structure as traditionally postulated. This is because business processes provide the means by which an organisation conducts its business and implement its strategies. They provide the framework for the deployment of people and other resources to meet the needs and wants of the customers. Customer satisfaction results from how effectively and efficiently an organisation executes its business processes.

From this process perspective, we look at IBM Malaysia as nothing more than a giant feedback loop that begins with identifying the needs of the customer, and ends with confirming satisfaction. The loop represents the composite of all customer-directed business processes, each aimed at delivering values expected of the customer.

We called this top level view the enterprise process. In our case, it identifies the core processes required to realise the objective of being the best at meeting customer needs and wants. A simplistic view of this is outlined in Figure 2. This is what the CEO of the organisation owns.

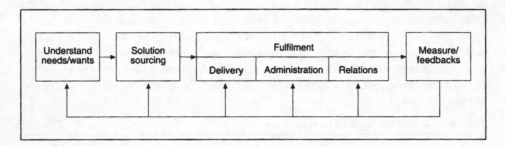

Figure 2   *Enterprise model*

The CEO then assigned each process execution responsibilities to his/her senior managers, thereby establishing responsibilities at the next level. Each senior executive in turn defines the quality objective he/she owns. In the same way, they will break down their individual processes into sub-process activities, relate these to a new set of process ownership, identify the process inter-dependencies, and define the relevant measurements.

Process improvements

Whenever inter-process dependencies are identified, process improvement teams (red teams) are formed to monitor deviation (defects) and take actions to correct and prevent defects.

Our process work has identified 95 customer driven processes with over 400 indicators and measurements across the whole organisation which encompasses the business we are in and the activities we do. They are essential to our success.

We found that the best way to cope with this magnitude in process works was to opt for a standardised process definition method, such that the sharing of experiences can be maximised and learning optimised. Our process management methodology (Figure 3) consists of three major phases, with each phase made up of several steps.

Phase 1: Process definition

This involves defining a process objective stated in terms of the customer value to be delivered. This is key since everything else about this process is targeted at realising this objective. These include the activity model that identifies task dependencies and requisites to successful performance, and the measurements that track impact and compliance. Some of the requisites are skills, tools and information.

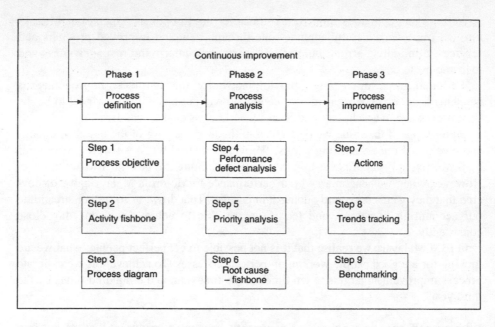

Figure 3 *Process management methodology*

We track compliance in terms of deviations from the activity model we define. As an example, if we determine that we want to do a tele-survey of our customers every 90 days to assess their satisfaction, then a percentage measure of customers who have not been called on after 90 days will alert us to the vitality of this process; whether it is working or not. We adopted the Motorola Sigma measure to promote improvements. Thus six sigma refers to three defects (deviation) per million operations in execution.

In our model, perfection is measured relative to the target for success (TFS) that we determine to succeed in the market place. TFS represents the specific design points of the process stated in objective terms for customer satisfaction and efficiency. It embodies our MDQ premise that quality is satisfying the customer efficiently.

Process quality then relates to the improvement required to get from where we currently are (referred to as the measurement baseline) to where we want to go (the target for success). This ensures that the cost of quality is synonymous with the cost of doing business successfully.

Phase 2: Process analysis

In this phase, data are collected to evaluate the performance of the various processes relative to the TFS. The defects are analysed, prioritised based on impact on the customer, and root causes are determined.

Phase 3: Process improvement

Since our business success depends on our ability to transform these processes into strategic capabilities that constantly provide superior value to our customers, it is of utmost importance that we adopt a systematic approach to process improvement.

We look on continuous improvement as referring to both incremental and break-through improvements; enhancing value through new and improved products and services: eliminating errors, defects and waste; and improving responsiveness and reducing cycle time.

We found that a key factor to spurring innovation for continuous improvement is benchmarking. Process improvement cannot do without this. Given our goal to be the best, we resolve to benchmark ourselves to world class quality standards.

An example of the value we derived from this was the case of the one-day, month-end closing of our accounting books. We were satisfied when we were closing our books in five days because at that rate we were among the best in IBM worldwide. However, when we benchmarked our performance to Motorola Malaysia, we discovered that they were consistently doing it in one day. That drove us to further streamline our accounting processes, and today we are able to achieve this one day close consistently.

In IBM Malaysia, we realise that it is not possible to get perfect people; what we are looking for are good people working in perfect processes. To achieve this we calibrate process improvement targets to ten times in the first year, and a hundred times by the third year.

With over 95 processes going on concurrently, it is most important that we put into place a number of mechanisms to ensure that they are all being executed and tracked. Many of these represent our own innovation and learning.

We invented the Process Vitality Index (PVI) to provide the means to assess the movements made in process works. The PVI monitors the progress made by each process against the three phases of our process management methodology, similar to the way we look at a particular composite index to assess movements in the stock market. Table 2 defines a sample PVI structure and shows the factors that go into the measurement of this index.

Table 2   Process Vitality Index

Process Building Phases	Points
A: Process Definition	
1. process/sub-process identified	10
2. activity fishbone/process map validated and documented	20
3. target for success determined	15
B: Process Analysis	
4. indicators in place/prioritisation done	10
5. measurements in place/baseline established	20
6. root cause analysis initiated	15
C: Process Improvement	
7. benchmarking	10

Today, the aggregate PVI for the 95 processes in IBM Malaysia is 85 per cent, meaning that on the average, all of our processes are at the stage of root cause analysis. Of course, some are ahead while others may be behind.

Process management also provides the framework for every one of our employees to participate in changing the way the organisation does business, that is, by being a member of process improvement teams. Allowing and motivating our employees to participate means inculcating a sense that everyone owns his/her piece of the business. The top-down approach described earlier enables employees to tie their roles and contributions to the achievement of the organisation's goals.

To date, over 87 per cent of our people are involved – applying this process improvement methodology to change the way they do their jobs day-to-day, making them more responsive to customer needs, more innovative.

The reason for MDQ is the customer, the reason for success is the employee. With our MDQ completely deployed through the processes, the future of the company is in the hands of the employees. Recognizing and rewarding the right behavior is critical for continued success. In many cases, we discovered that existing award and compensation programs had to be revamped and strengthened in order to reinforce the desired behavior. As an example, a simple suggestion award worth RM20 for an idea yielded hundreds of new ideas on how existing processes can be streamlined

The role of senior management

A critical catalyst in enabling us to sustain the MDQ initiative has been the role played by the senior executives of IBM Malaysia. The MDQ transformation took hold because senior executives champion MDQ and model the way. Senior managers are involved in all aspects of MDQ.

We determined very early in the game that the best way to demonstrate this was to learn about quality ourselves, and then train others in order to cultivate the confidence that is necessary to implement a transformation like MDQ that has a course of five to seven years to reach completion. Since then, senior executives have attended teach-the-trainer (T3) classes as well as taught classes for employees in managing change – often referred to as the 'Its My Business' (IMB) series in IBM – empowerment, risk management, process management and assessment, and behavior paradigms.

Senior executives also lead in building a new IBM Malaysia organisation that is a composite of all the customer directed processes. Relating process to the organisation helps us identify opportunities to streamline and eliminate hierarchical layers. In doing this, we make sure we avoid duplication and that only identified customer values are serviced. As a result, we have reduced our management layers in several areas: in client relationship management from five to three, technical services operation from six to three, and customer administration fulfillment from four to three.

Process councils aligned to each of our business goals replaced management committees. A senior executive heads each process council and also participates in others. They select relevant processes from among the 95 registered to date for review.

Each process owner is expected to review their process measurement trends, the analysis of root causes that they had worked out with their red teams, and outline the support they require form the councils to carry through their action plans.

The councils validate, where appropriate, that inter-process relationships are present and measured. They also receive and approve recommendations for MDQ process awards.

A key role of senior executives in the councils is to look for common learning and issues that they feel may impact others processes. These topics are then brought to the relevant monthly leadership sessions for dialogue, and agreement to follow-up actions by the process owners and team leaders.

The two leadership sessions, a process leadership dialogue (incorporating all process owners) and a people leadership dialogue (incorporating team leaders), are in actual fact reformatted managers meetings. The sessions focus on validating inter-process relationships, verifying, where appropriate, that they are present and measured.

We also revamped our business planning model to align it with the enterprise business process model. The yearly commitment planning becomes our MDQ plan. It represents the things we need to do to reach the next destination in our MDQ journey.

Conclusion

In conclusion, from a competitive standpoint, the full integration of MDQ into our business ensures that our competitors cannot easily imitate our strengths. This gives our customers capabilities and values that they can get nowhere else.

The improvements in customer satisfaction, the continued growth of our business, and the quality awards we have received to date are testimonies to this commitment.

References

Brown, Paul L *Managing Behavior on the Job*, John Wiley & Sons, New York, NY,1982.

This case study was prepared by Rodzlan A Baker, IBM Malaysia, and was first published in the *Asia Pacific Journal of Quality Management*, MCB University Press, vol 3 no 3, 1994, pp 54–61.

Case study questions

1 Discuss the usefulness of conducting employee surveys as part of IBM Malaysia's quality improvement program.

2 Evaluate the process improvement methodology developed by IBM Malaysia in terms of the steps involved and their implementation.

C9

The Road to Quality: Currency Department, Reserve Bank of New Zealand

In the mid-eighties the New Zealand Government began a process of reform in the public sector. This reform had a profound effect on how government organisations were managed with the introduction of commercial criteria and objectives, as well as staff level reduction. At the same time, products and services were expected to be maintained and even improved. The Reserve Bank was not exempt from these reforms and, at an early stage, we (the Currency Department) decided to take the necessary initiatives ourselves. Staff levels were significantly reduced as the early focus was on efficiency.

It was obvious that morale within the Currency Department was poor. We were not focused on quality, and productivity and customer service was low on our list of priorities. The efficiency drive we had experienced in the early years had taken its toll. In deciding the 'best way forward' it was concluded that the conventional western management style of autocratic control, mistrust, conflict, and competition with peers should be challenged and a new approach tried. It was time to place our priority on effectiveness – how well are we doing our jobs, how could we do them better. The TQM core values of trust, empowerment, cooperation, harmony, and responsiveness appealed to us as essential ingredients to improve morale and our effectiveness. Thus, we launched our quality improvement program to challenge and change our traditional practices.

In the introductory phase we created quality teams in each office. Everyone was made a member of a quality team, participation was considered 'part of the job' as we wanted everyone to experience TQM and judge its value for themselves. We discussed the concepts of continuous improvement and the identification of work process related problems. A structured problem solving process was developed and introduced and training provided for team leaders.

Our teams started to meet for an hour each week to review work processes, identify process related problems and develop and introduce solutions. Initially there was much scepticism – was this another management fad that would pass quickly, was this another efficiency measure? However as we all gained a better understanding and experience of the TQM concept, we became more confident and began tackling more complex problems.

Our organisation culture began to show signs of change for the better as trust replaced conflict and suspicion, and our top-down communication was replaced by two-way communication. Management was listening to staff and staff to management as we all become actively involved in the decision making process. We were beginning to develop a quality culture with benefits for us all.

The next step down the quality road was the introduction of self-directed work teams. We wanted to extend quality management into all aspects of our work and our teams were given the responsibility, and its associated accountability for planning and organising work flows, assignment of tasks, utilisation of resources, and monitoring performance, progress and achievement. Our teams are encouraged to work together in consensus planning and decision making to achieve shared goals and objectives. They are also very much focused on identifying work process improvements as part of our strong emphasis on continuous improvement.

Three years later with our TQM programme established and working well it was decided that the time had come to carry out a review of all of our work systems and processes.

It was decided that our focus should be quality systems and processes which provided a level of control that would dramatically reduce errors and the need for rework. We moved away from the traditional idea of dual control and checking focusing on responsibility and accountability. This has resulted in people having more pride in their work with a huge reduction in errors.

As part of the annual planning round each year we set a number of performance criteria under the headings of quality, quantity, timeliness, efficiency and effectiveness. These criteria are set after a consultation process involving everyone in currency processing. We closely monitor our progress and achievement in terms of these performance criteria. Work process problems which are inhibiting achievement of high levels of performance, are identified as part of the monitoring process. Teams will then take action to solve problems and improve performance. On-going reviews of performance promote commitment and motivation and allow us to all share in the achievements of goals and objectives.

Currency Operations has a huge number of customers – everyone who uses notes and coins. As part of our drive for quality we commissioned a research organisation to conduct a survey to ascertain the needs and expectations of the public in regard to the currency in circulation. The research organisation provided us with an extensive report which provided a great deal of information to help us improve our products and services.

Our direct customers are the commercial banks and although we are the monopoly supplier of our products in this country (it is illegal for anyone else to produce currency) it was decided that we should set up a customer focus programme to look at the ways cash is handled, to ascertain the needs and expectations of the banks and to improve communication and our working relationship. By working with our customers and getting a greater understanding of their needs and expectations we can improve the service we provide to them. To date, customer response has been very positive.

ISO 9002 is an international quality standard and it was decided about a year ago that we should work towards gaining accreditation to this standard. It sets out a number of quality focused criteria covering all aspects of a quality system, irrespective of the product or service offered. An organisation seeking ISO accreditation, documents its systems and procedures in a quality manual and these systems and procedures are then measured against the requirements set out in ISO 9002. The quality documentation is a means of showing and coordinating the relationships, responsibilities and activities of personnel and the functions of the quality system.

Currency Operations has recently been accredited with ISO 9002. Gaining this accreditation involved having Standards New Zealand audit our systems and procedures to confirm our compliance with the standard. Accreditation confirms we meet the international quality standards and demonstrates our commitment to quality.

What have we achieved over the past few years? Has our commitment to Quality and TQM been worthwhile? We have established a very strong focus on quality and are now very much quality-driven. As a result of this we have achieved considerable gains in productivity. Perhaps more importantly, we have changed our organisational culture. The team concept is now firmly established in Currency Department and everyone is involved in managing quality. We are all involved in the decision-making process and there is wide consultation and sharing of information. Many work process problems have been solved and there is a focus on continuous improvement. Trust and communication are greatly improved and anyone can express a point of view without fear or ridicule. We are committed to quality and work together to meet the needs of our customers.

I have heard it said, 'Quality is an attitude' and I think that this attitude is now an integral part of the way we do things in the Currency Department. However the quality journey has not finished for us and we continue to find ways to improve and to make Currency Department a better place to work.

This case study was prepared by Gary Wilmshurst and Brian Lang, Currency Department, Reserve Bank of New Zealand.

Case study questions

1 What are the key features of implementing TQM in a service organisation such as the Currency Department of the Reserve Bank of New Zealand?

2 Discuss the challenges faced by a service organisation in seeking ISO certification.

C10
Continuous Improvements in New Zealand Ports

Introduction

In 1988 New Zealand ports ranked amongst the worst in the developed world for productivity and profitability. Five years later the international shipping community were rating New Zealand ports as among the best in the world for turnaround of vessels and cost effectiveness. In addition, the New Zealand ports in general are providing some of the best returns to shareholders of any industry grouping in the country. How has this remarkable transformation been achieved, and what were the specific forces driving the improvement process?

This case study traces transformation of New Zealand ports and examines the ongoing push for the continuous quality improvement. One New Zealand port that has excelled in the service quality area is the Port of Timaru. Amongst the first four ports in New Zealand to gain Port Company status, the Port of Timaru has gone on to post a number of notable firsts and has recently become the first port in New Zealand and Australia to gain ISO 9002 Certification for port operations, namely its Freight Station and Container Yard operations.

The Port of Timaru, is used as a case study upon which to draw out the specific changes which have propelled New Zealand ports into top international ranking for all performance measures.

The case examines the 'third phase' of changes occurring now in New Zealand ports, and which reflects a maturing of the change process towards a concentration on:

- continuous improvement of port services
- developing employee commitment
- exceeding customer expectations
- broadening skills/responsibility.
- Innovation and technological development.

New Zealand ports are now concentrating on locking in the gains of the past five years and many are pursuing ISO 9000 Series Certification. As more independent cargo handling transport operators gain ISO Certification 'quality service chains' are developing around the country linking New Zealand exporters to road and rail transport operators, to ports, and shipping companies, in an unbroken quality chain.

As GATT brings down tariff barriers to New Zealand exporters, it is feared that other barriers will go up in their place. New Zealand exporters are determined that quality assurance will not become a barrier to successful worldwide trading.

Background

Up until the late 1980s, the productivity and profitability of New Zealand ports was one of the worst in the developed world. Ports, by enactment of Parliament, were run by publicly elected harbour boards. They had a wide range of responsibilities including commercial port operations, recreational activities, regulatory functions, as well as non port-related commercial undertakings. These conflicting responsibilities and lack of clear commercial focus, combined with a monopolistic system of waterfront labour provisions, led to poorly performing, inefficient ports. Return on investment in New Zealand ports was a dismal 2–3 per cent and capital investment was poorly directed (Prebble, 1993).

The structural problems established by statute were compounded by the monopoly granted to unions to cover particular work areas, and by the pooled labour system of waterside employment. Restrictive work practices, over-manning, on the job wage bargaining, demarcation disputes, and spelling were just some of the baggage carried by ports to 1988. This case study describes the changes that have taken place in New Zealand ports, in particular the improvements that have been achieved at the Port of Timaru Ltd.

Phase One: Structural reform

In March 1987 the New Zealand Government set the following objectives for structural reform of New Zealand's ports:

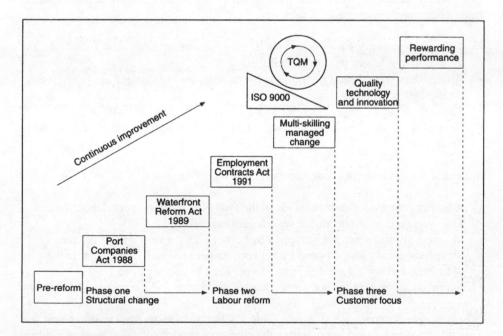

Figure 1 *Three phases of continuous improvement*

- The separation of commercial port activities from non-trading roles.
- The removal of legislative controls over commercial port activities.
- The introduction of private sector business standards of accountability to ports.

Structural reform was a major driving force in New Zealand ports, pursuit of productivity, progress and profits, and could be considered as phase one of a continuous improvement journey as represented in Figure 1. The two main pillars of structural reform were:

- the Port Companies Act 1988; and
- the Waterfront Reform Act 1989.

The Port Companies Act required each harbour board to form a port company under the New Zealand Companies Act, then pass the commercial port operations over to that company. The key results of this change were:

- Ports had strictly commercial goals.
- The harbour boards monopoly on the provision of cargo handling plant was repealed.
- The national control of port planning under the New Zealand Ports Authority was removed.
- Ports were required to run as 'successful businesses', to pay company tax, and provide dividends to their new shareholders.

The Waterfront Reform Act followed six months later, to reform the structure of waterfront labour employment. Key results were:

- Waterfront Industry Commission (WIC), a government instrument for controlling the waterfront labour pool, was abolished.
- Watersiders were directly employed by stevedoring companies, including newly established port companies.
- The waterfront no longer operated under any special labour legislation.
- New port codes of employment were negotiated to replace the National Union Award.

Structural reform brought about many benefits, including:

- A 50 per cent reduction nationwide in the total New Zealand ports labour force.
- The average time alongside for vessels also halved.
- Manning strengths per gang dropped by between 25 per cent and 75 per cent.
- Stevedoring costs also dropped by between 25 per cent and 75 per cent. In the case of loading livestock, rates fell from 75c per head to 25c per head.
- Return on investment for port companies had risen from 2–3 per cent to 13.5 per cent.

Other changes occurring are shown in Table 1. Clearly, this structural reform phase was a significant step forward in improving the quality of service provided by New Zealand ports.

Table 1 Changes achieved from structural reform

	Pre-1988	1991
Port development	Nationally controlled by NZ Ports Authority	Determined by each port on a commercial basis
Watersider employment	Nationally controlled by the NZ Waterfront Industry Commission (a pool system)	Locally determined direct employment by companies
Watersider conditions	National awards	Local schedules
Manning	Manning levels set in national documents (3,000 + watersiders)	No manning scales in local contracts (<1,500 watersiders)
Restrictive work practices	Widespread (on-the-job bargaining, spelling, inflexible, work practices)	Removed (no second tier bargaining, flexible skilled workforce)
Labour utilization	Low utilization, high gang strengths	High utilization, low gang strengths
Casual labour	None	Widespread use of casual workforce
Provision of cargo-handling plant	Monopoly at each port by Harbour Board	Free competition
Work hours	Primarily first and second period (0700-2300 hours) Monday to Saturday, excluding public holidays	24 hours per day, 365 days per year
Technology	Barriers to the introduction of new technology	Rush to embrace new technology in a more competitive environment
ROI	2-3 per cent	13.5 per cent

Phase Two: National labour reform

To increase the competitiveness of the New Zealand economy and encourage more rapid change, the Government introduced a fresh approach to labour relations in the form of the Employment Contracts Act 1991 (ECA). The act replaced the more traditional Industrial Relations Act with legislation which put employment agreements on the same footing as other commercial contracts between parties. It provided employers with the ability to enter into individual contracts with workers previously covered under Union monopoly bargaining. The Act enabled employers to 'lockout' workers in pursuit of a renegotiated contract. It stripped away the concepts of demarcation and compulsory union membership.

With demarcation gone, multiskilling flourished and many of the remaining restrictive work practices vanished. The changes that national labour reform provided are shown in Table 2. The changes process initially had a win-lose flavour and led to resentment and ill feeling by organised labour groups at many ports, primarily due to the high level of job losses.

Table 2 Changes achieved from national labour reform

	Pre-1991	1993
Employment conditions	National union based awards with local port schedules	Local employment contracts (under the ECA 1991)
Bargaining	Collective bargaining	Individual bargaining; collective bargaining
Work coverage	Exclusive demarcation coverage area for each union	No exclusive coverage; Multi-skilling flourishing
Company culture	National negotiations over awards led to a uniformity of conditions and stunted the development of individual company cultures	Local port negotiations led to a rapid diversity of contracts and the emergence of quite different company cultures
Individualism	National awards tended to average employment conditions to the detriment of high performers	Individual contracts have provided scope for setting conditions specific to individuals particularly those in key positions, and high performers
Management	Management's hands were often tied and change was slow	No excuse for not managing change in an intensely competitive environment

Over time the positive aspects also became apparent to waterfront staff, and these included:

- more job satisfaction
- less idle time and boredom
- higher wages (though lower effective hourly rate)
- introduction of new technology
- better management of capital expenditure
- better allocation of resources
- more professional and focused management.

The continuous improvement cycle for New Zealand as a whole, and New Zealand ports in particular, received a timely boost from the Employment Contracts Act. It was apparent that the change process was accelerating as competitive forces worldwide and within New Zealand increased.

Phase Three: Customer focus

The maturing of New Zealand ports continued during 1992 and 1993 with the development of a stronger focus on the customer. Attention was diverted from an internal focus on port efficiencies to an external focus on customer expectations. Total Quality

Management, benchmarking performance, ISO 9000 Certification, skill building, recognising and rewarding performance, and quality technology and innovation all became important issues. Ports became more aggressive in seeking out from customers those key factors which were important to them. Key indicators such as vessel schedule maintenance were monitored. Other key indicators are outlined later in this case under the heading 'Information and Analysis'.

Ports began to really focus on quality of service, and on continuous improvement of service quality. New improved systems and procedures were designed and implemented. New and modern equipment has been installed and ports are developing at an increasing rate. At the forefront of these changes have been the North Island ports of Tauranga and Napier, and the South Island ports of Nelson and Timaru. The average improvement in productivity and profits for all New Zealand ports for the period 1988–93 are shown in Figure 2.

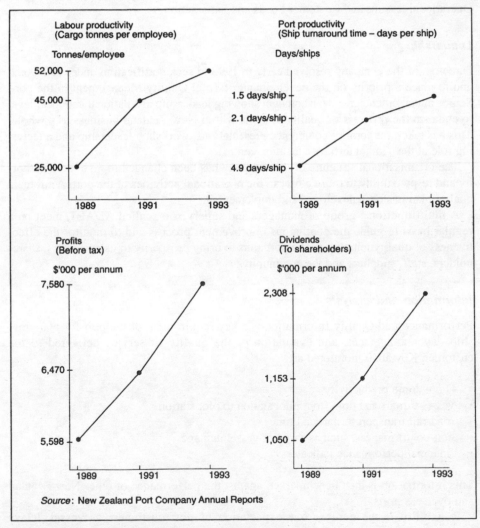

Figure 2 *New Zealand Ports: Performance measures*

This customer focus has led to spinoffs for the port workforce as ports come to the realisation that the next level of improvement can only occur with a well-trained, motivated and committed workforce. New Zealand ports are currently at differing positions with the move to empowering that workforce to build quality into the service they offer.

Port of Timaru Ltd

The following sections describe the steps taken by the Port of Timaru Ltd to break through to a new level of customer service and port performance. The description follows the headings adopted in the Malcolm Baldrige Quality Award Criteria, namely: Leadership, Information and Analysis, Strategic Quality Planning, Human Resource Development and Management, Management of Process Quality, Quality and Operational Results and Customer Focus and Satisfaction.

Leadership

Directors of the company resolved early in 1992 to seek certification under ISO 9002, and to place a priority on the development of Total Quality Management in the port. Senior management have been active in showing leadership at a national and local level to enhance the progress of quality systems within New Zealand business as a whole. This has taken the form of conference presentations, workshop leadership and a lecturing role at the Timaru tertiary education centre.

The organisational structure of the company has been changed to give more authority and responsibility to those closer to the operational activities of the port. A strategic quality plan has been developed and implemented.

A multifunctional group of managers and supervisors, called ADAPT, meet on a regular basis to guide the continuous improvement process and to monitor the effectiveness of quality initiatives. The port aims to bring prosperity to its customers, shareholders, staff, suppliers and the community.

Information and analysis

Performance and Quality Information is a key requirement of the port for planning, ship day management, and evaluation of the quality of service delivered to the customer. Key areas monitored are:

- shore crane productivity
- vessel turnaround time from pilot station to plot station
- road/rail transport turnaround time
- non-conformances, such as cargo or vessel damage
- financial performance indicators.

This information is then benchmarked against the performance of other New Zealand and overseas ports.

Reliability is the number one expectation of port customers as vessel delays can become very costly. The criteria used to select key performance measures were determined at customer interviews where they were asked what were the most important indicators of port performance.

Strategic quality planning

Short-term (one to two years) strategic planning for quality at the Port of Timaru is embodied in the company's Strategic Quality Plan. The plan is implemented under the leadership of ADAPT, referred to earlier. Longer term (greater than two years) strategic planning is carried out in the business planning process under the leadership of the port directors and management.

Recent changes to the organisational structure have led to the total integration of quality management into the daily activities of all process owners (managers and supervisors). Process owners have been empowered with the authority to design and implement quality systems within their own areas. The future survival and prosperity of the port hinges on strategic initiatives producing results.

The past two years have seen an improved performance at the port in a number of key areas:

- turnover is up despite reduced charges
- profitability is up
- cargo throughput is up
- vessels handled are up
- service quality is up.

Human resource development and management

The Port of Timaru Business Plan contains HRM objectives aimed at continuous improvement, such as:

- stronger links between performance and remuneration
- self-regulated health and safety organisation
- broadened areas of responsibility and decision making
- utilisation of information on new technologies, processes and attitudes
- increased training and education
- better allocation of labour.

These objectives are broken down into smaller goals and tasks with timelines for completion.

Staff involvement occurs in a variety of ways including:

- performance improvement teams (six currently)
- health and safety committee
- ADAPT.

Extensive time is given for consultation and consensus forming during employment contract negotiations. The company's workforce is flexible across departmental boundaries leading to better utilisation and a need for more effective training. A total of 21 operating certificates are now available for staff upon completion of internal port company training.

Management of process quality

Traditionally the quality of service processes has been maintained by the close way in which the Port of Timaru Ltd works with its suppliers and customers. This has been supplemented at the cargo interface by ISO 9002 certification. Procedures are now well defined and documented, staff training has become more structured and complete, and Corrective Action Requests provide an avenue for recording non-conformances and for ensuring they are dealt with. Company-wide ISO certification is expected within eight months.

Supplier quality is an area where more work is required to ensure the total transport chain from New Zealand exporter to overseas customer exceeds the expectations of the customer. The port company is currently assisting a number of suppliers in their quest to implement continuous improvement programmes and gain ISO certification.

A recent customer survey indicated that the strengths of the Port of Timaru were in vessel turnaround time, industrial relations, vehicle access, staff attitude and management contract with customers. Weaknesses identified included port charges, documentation, and transport costs to and from the port. Since 1991 the port has worked hard at maintaining service quality in the areas of strength and increasing service quality in areas of weakness.

Major restructuring of all activities has taken place, multiskilling of staff has been successfully accomplished, and equipment modernised in an effort to reduce port charges and increase service quality. The port has entered into various strategic alliances with transport operators to improve backloading potential and bring down transport costs to and from the port.

Quality and operational results

To quote one international shipping line customer currently calling at the Port of Timaru and four other New Zealand ports:

Port Time. In 1988 our line spent ten days in four New Zealand ports on each voyage. In 1994 we are spending seven days in five New Zealand ports. This is caused by being able to work 24 hours per day and increased productivity.

Productivity. This has increased 97 per cent in New Zealand ports for our line on a cargo handled per day basis, since 1988.

Costs. Cargo handling costs have been reduced by about 40 per cent to 50 per cent. Vessel costs have not reduced in some ports. However, I cannot go around what Timaru has achieved in terms of savings (over 30 per cent reduction in vessel costs).

Far East port and cargo costs have increased by about 25 per cent (over 45 per cent in Japan) in the period 1988–94 for our shipping line.

Overall. The reforms have meant a lot to our shipping line. In terms of schedule maintenance we used to rely on the flexibility of the Far East. Now we rely heavily on New Zealand in order to make up lost time. Port companies are now commercial entities which has resulted in less cargo damage, cost savings, more focus on mutual customers (shippers, consignees) and a better service all round.

Clearly, the Port of Timaru Ltd, and New Zealand ports in general, have improved markedly. However there is always more that can be done.

Customer focus and satisfaction

Another recent Port of Timaru customer survey was targeted at the shipper group who were most critical in the areas of port charges, documentation and land transport costs. The responses were overwhelmingly positive with 67 per cent rating the Port of Timaru's performance as 'excellent' while the remainder rated it as 'good'. None of the respondents rated the port's performance as less than 'good'.

Conclusion

The port has journeyed far from the attitudes and practices in existence during the 1980s. In April 1994 the port was successful in winning the top Silver Pyramid Award in the Aorangi Region Business Development Quality Awards. In presenting the award the New Zealand Government Minister of Business Development, Roger Maxwell, said the company had shown excellent progress in all seven (Malcolm Baldrige) assessment criteria and had clearly identified their position in a rapidly changing industry with wide employee involvement and high customer focus.

References

Prebble, R, 'The New Zealand Experience', Ports Conference, Tauranga, NZ, January 1993.

This case study was prepared by R J Weaver, Port Timaru, New Zealand, and was first published in the *Asia Pacific Journal of Quality Management*, MCB University Press, vol 3, no 3, 1994, pp 45–53.

Case study questions

1 Discuss how the structural and labour reforms in New Zealand ports helped to improve the performance of the port of Timaru.

2 What role did the senior management team of the port of Timaru play in achieving customer focus?

3 Identify and evaluate the restructuring which has taken place at the port of Timaru.

C11
The Sony Precision Engineering Center (Singapore)

Introduction

Despite being the smallest nation in South-East Asia, the economy of Singapore is one of the most prosperous and its people enjoy the highest standard of living in Asia after Japan.

The economy has been carefully orchestrated as a series of deliberately phased movements, for example the initial low labour costs of the 1950s were used to attract investment opportunities, then during the 1970s Singapore's geographic position and transportation links were used to attract industry.

Even today Singapore is using its competitive advantage to ensure investors will use Singapore as a high-technology base to design, manufacture and distribute products which are made in low cost factories in a growth triangle within a 50 km radius of Singapore (including parts of Malaysia and Indonesia). This shift of labour intensive industries from Singapore has paved the way for the entry of high technology industries, including the Sony Corporation of Japan.

With its highly valued education base, excellent transportation and communication links and a work ethic that has been strongly advocated, the industrious people of Singapore were ideal candidates for the establishment of Japanese companies within their midst. This is not to say that Singaporeans had to forgo their ethnic identity or way of life. Far from it. As one of the Sony employees in Singapore commented 'We will learn from the Japanese, but we will do it better and we will still enjoy the Singapore way of life.'

With these attributes and the constant pursuit of improvement, it is not surprising that the Sony Corporation chose to locate not one but six of its facilities in Singapore.

Corporate overview

The Sony Precision Engineering Center (SPEC), is a wholly owned subsidiary of Sony Corporation of Japan and as such is said to carry the motivational power of the 'Sony Spirit'.

Although this spirit can not be felt or touched it can be best described as a culture of pride which allows for the personal contribution by each employee. This aspect of SPEC will be expanded upon later in this case study, but for the record Sony Chairman Akio Morita says that creativity comes not from machines or computers. 'It requires human thought, spontaneous intuition and a lot of courage.' Freedom to discuss, freedom to experiment, freedom to innovate are the keys to creativity and basis of the 'Sony Spirit'.

SPEC itself was founded in Singapore in 1987 and comprises six factories on two sites, supporting Sony operations in Asia through high-precision manufacture of component parts for compact disc players and video cassette recorders. Many of its products are also exported back to the parent company, Sony Japan. SPEC also maintains its own factory automation division within the R & D department located in Singapore's Science Park.

Within just two years after SPEC began operations in Singapore value added productivity was already 2.6 times the industry average and its profitability 1.3 times the industry average.

SPEC's mission statement (see Appendix 1) has three main objectives: to manufacture precision components; to provide specialist high-tech services in factory automation; and to act as a technological support base for Sony operations in the region. It emphasises only one vision however – to be a centre for excellence in all its activities.

SPEC utilises a great amount of 'home grown' innovative technology in much of its manufacturing and assembly operations. As an example the company has developed a remarkable automated production line for optical pickup units, which is the single most important component of a compact disc player. This process was the first of its kind outside Japan and marks a major achievement in robotic technology and localised parts supply. The output of optical devices from this process, which is 90 per cent automated, represents one third of Sony's worldwide production or 20 per cent of its international share. Although this technology is 'state of the art' the equipment was primarily used as a means of assisting the worker rather than as a form of replacement for labour.

Acrylic lenses are manufactured in a single, direct moulding procedure which had only recently been developed. Again these lenses are an integral part of the compact disc optical drive requiring precision accuracy in a clean room environment. This lens operation at SPEC is Sony's pioneer and the only high-quality lens production facility.

Video-head tips are produced by SPEC's Magnetic Head Division which was the first Sony plant outside Japan and has the capacity to produce one million video tapes per month. The division supplies video tips internally to SPEC's Video Drum Division and exports them to customers in Japan and other countries.

The cylinder or drum is the key component of the VCR and is assembled in a process which is 90 per cent automated using integrated technology backed by highly trained operators. The cylinder is assembled from high precision parts. The accuracy of machining is controlled with up-to-date CNC technology and the most precise measuring instruments.

SPEC also manufactures power supply units (PSU's) for original equipment manufacturers in the USA, Europe and ASEAN. Much of the design responsibility and vendor sourcing came from Japan, however SPEC aims to concentrate on localised sourcing and design, releasing more funds for R&D and more automation. An interesting feature of the assembly process is the 'odd-shaped' insertion machine developed by Sony. The six stations, all manufactured by SPEC's own factory automation division, form the only line of its type in the world.

A fully automated moulding process manufactures parts for use in compact disc players, optical pick up devices and VHS video cassettes, with exports to many countries around the world.

The machining department produces many components including tape guide pins for VHS and Beta VCRs. A striking feature of the machining department was housekeeping and cleanliness, despite the potential dirt and contamination associated with machining.

Human resource management issues at SPEC

When trying to understand what made SPEC 'tick', it was obvious that a sense of commitment and pride was exuded from everyone. Everyone had a shared meaning and understanding as though they were all going in the same direction.

Why was this? For one thing, there was trust, a two-way trust. A trust that broke down barriers between so-called workers and management, a trust that recognised the dignity of, and thus respected, the individual and their contribution to the welfare of the organisation.

A positive environment has been created at the workplace where all people are actively encouraged to make a contribution, where positive feedback is given freely and equally, and negative constructive feedback is given and received appropriately. What can be learned from SPEC is that rewards and bonuses are given for excellent achievement and the organisation is willing to invest in innovative ideas of employees. This does not imply that SPEC have endorsed paid employee suggestion schemes. As one employee of SPEC commented 'It is not done here. What is expected of the employee is to have an attitude to want to do 100 per cent right. It is part of the job. Improvement of the manufacturing product is not a choice.'

Teamwork

SPEC is characterised by teamwork and people orientation is visible everywhere in the factory. From charts in the assembly area updated by operators, pictures at work stations to, what must surely be one of the strongest indicators of teamwork, a board in the entrance foyer signed by approximately 1000 employees pledging their commitment to SPEC on the company's third anniversary of operation.

The role of each team member at SPEC is to manufacture the parts, ensuring quality is maintained at the correct specification, and then when minor defects do occur they are attended to by the operator. Minor breakdowns are also handled by operators with trained technicians and engineers only assisting if the task is too difficult. This contrasted quite markedly with companies in the West, using similar equipment but with less emphasis on employee involvement and process ownership by the workers. SPEC employees worked in teams to the extent that they were responsible for all facets of the process including material handling and cleaning requirements. This resulted in no visible inventory on the shop floor and a shop floor that was exceptionally clean, particularly in traditionally dirty areas such as the Machining Department for producing video tape guide pins using centreless grinding machines and milling machines – no swarf, no oil, almost a 'clean room' environment.

Given that SPEC place a great deal of emphasis on building a standardised design across its product range where practicable, this has led to minimal job change or change-over setting time. Where setting time is required it is simply a matter of trimming and adjustment of standard products for different applications. What SPEC have achieved is an operator team process that has been designed with robust parameters requiring minimal change or modification and using products that have been designed with standardisation in mind.

Team members also assist in training other team members. All operators are trained in how to read charts (process control, targets, etc).

The discussion so far relating to teamwork has focused on the shopfloor manufacturing and assembly areas where teams tended to be both cross functional and product based, however it should be noted that teamwork applied across the site, including service and administration departments. All teams are eligible to participate in the Productivity Campaigns discussed later.

Training

New employees receive 'on the job' training and all new shop floor employees are quality trained. Once familiar with their particular workcell, cross training and new skill development takes place. By far the biggest emphasis on training was that of quality training which is covered in detail in a training plan.

An important feature of the SPEC quality training program is the concept of CEDAC (cause and effect diagram with additional card). This is used extensively throughout SPEC by the continuous improvement teams who after 'brainstorming' possible solutions to problems use a colour coding system to monitor the effectiveness of the fixes made. This provides for a readily visible performance measure of each team's ability to solve problems.

The first person to be trained was the CEO and he now forms part of a management audit team that conducts monthly audits of all departments using CEDAC. In other words a classic MBWA (Management By Walking Around) approach. CEDAC allows all employees to play a role in identifying and solving problems and therefore making a direct contribution to both quality assurance and productivity.

Training has proved to be a vital factor in the success of SPEC. In addition to structured in-house programs for new employees, SPEC runs courses on industrial automation, quality management and information technology, conducted by internal and external trainers.

Unions

In Singapore, unions are recognised by individual companies but not through government policy. People tend to join for social benefits such as discount cards, and not for job security. Japanese companies (Sony and SPEC included) tend to operate a lifetime employment policy which in Singapore may not be so necessary (on the basis that jobs are plentiful in Singapore). The union at SPEC is generally involved in collective bargaining covering such issues as health and dental benefits and meal allowances.

Productivity Activist Committee (PAC)

As part of the continuous improvement activity at SPEC, a committee has been established to monitor and reward the high achievers within the organisation. Periodically this group hosts a 'Productivity Month' campaign to further enhance the already successful employee participation programs. In conjunction with the campaign, a productivity presentation competition is held where competing teams vie for the title of champion productivity team. An extension of the 'Productivity Month' is 'The 100 per cent club'. This club has been set up with the intention of promoting the 100 per cent performance concept and also to give recognition to employees who have constantly achieved 100 per cent in one or more areas of their work.

Nominations for club membership are reviewed by the PAC ensuring that the established criteria for the 100 per cent club are met, including matters such as work performance, attitudes, attendance rate, communication skill and leadership quality. An interview is conducted by a 100 per cent club sub-committee to gauge and evaluate the nominees in terms of their awareness of the company activities and work performance. Another interview is then conducted to determine first, second and third placings. All nominees of SPEC's 100 per cent club receive a letter of certification and a souvenir for their outstanding performance. Cash vouchers are also awarded to the top placings. This competitive activity appears to bode well for continued employee participation in the ongoing welfare of the SPEC organisation.

More people issues

Apart from the attraction of a long life employment system (if they so desire), SPEC employees also enjoy the benefits of a direct relationship between their performance and pay. Bonuses are paid for productivity and quality improvements that lead to greater profitability for the organisation. These bonuses equate to a maximum of one month's pay on one occasion per year with an additional one month bonus for individual performance. It seems a simple philosophy – if you work harder and smarter you are rewarded and at the same time the company benefits through the improvements made by the team or individual.

In addition to providing a wage element based on a sound level with realistic incentives, the young enthusiastic management at SPEC provide constant motivation to its team, encouraging SPEC employees to produce good results and contribute to the improvement of productivity, which in turn leads to a rise in their own material standard of living.

The combined effort of SPEC's employees is not contained only within the boundaries of its factories. In true Japanese ideology, Sony and SPEC believe it is important to nurture the needs of the market which it serves in terms of maintaining the welfare of the community. As an example, more than 70 per cent of all SPEC employees contribute to a community chest charity program. For each employee contribution made the company contributes an equal amount.

Status differentials did not seem to be so apparent at SPEC. It was obvious that everybody wore the same regulation grey Sony tunic or jacket, whether they were management or process worker or service department worker. Offices tended to be open plan, again breaking down the barriers between management and workers, the only exception being where customers are 'entertained' or interviews held. Common canteen facilities are provided, subsidised by the company and catering to the needs of the predominant ethnic group.

Quality assurance

One can not walk away from SPEC without first witnessing the incredible display of teamwork and sense of commitment and enthusiasm shown by all the employees. Secondly, but possibly just as crucial to all at SPEC was the relentless pursuit of perfection in everything they do. In fact, the Sony Quality Policy sums up the objectives of SPEC more than adequately:

- make Sony the standard of quality worldwide
- establish a competent Quality Management program in every operation, both manufacturing and service
- eliminate surprise nonconformance problems
- reduce the cost of quality

This policy is driven by what SPEC refer to as 'QM means Communication'. The policy is communicated through both the operations and systems side of the business with individual roles and accountability enforced each step of the way. One of the SPEC Quality Management programs is called Quality Management by Determination, Education and Implementation (QMDEI) and is a top down approach which demonstrates management's commitment to quality improvement and passes this sense of commitment down the line through policies, operational procedures and systems.

SQIM is a very important feature of the quality program at SPEC. SQIM means SPEC Quality Improvement Month which is an exercise designed to motivate all employees to become involved in the quality improvement activities. An integral part of the campaign is a physical measurement method that helps define the best quality performing production department. SQIM is held every six months and includes such criteria as customer return ratios, line performance indexes and quality improvement presentations.

The contents of SPEC's Zero Defect (ZD), and quality improvement activity can be best summarised by reviewing Figure 1. Briefly the ZD elements relate to customer satisfaction, statistical quality control, problem solving, employee participation and quality management.

An interesting approach to continuous improvement and problem solving is what Sony refer to as Window Analysis (window of opportunity perhaps). Here, a problem is first acknowledged and then through a series of steps it is analysed taking into account the degree of knowledge possessed by employees, number of employees involved and whether training is required. It addresses issues related to communication, adherence and development problems.

The impression formed was that SPEC's customers are involved each step of the way where possible. SPEC conduct intensive training programs for its major customers as well as maintaining a monthly liaison meeting schedule. Handbooks are made available to each customer which detail a product return procedure and correct handling procedure, thereby making it easier for customers to register their complaints and providing improvement opportunities for SPEC.

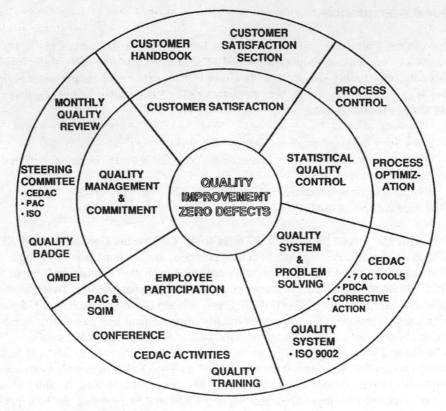

Figure 1   *SPEC's ZD Elements*

Evaluation

The lessons to be learnt from Sony and perhaps other Asian companies are simple. So simple that perhaps we will never understand these fully. So simple that we have seen it all before and cannot reason why it is applied so successfully by the Asian companies. These are successful companies that treat their employees with respect and dignity and recognise the significance of each job so that each job will produce quality output.

SPEC have created an environment where its employees want to work for the benefit of the organization, want to contribute to its welfare and maintain its position in the marketplace but also within its community. It has recognised that manufacturing, or more appropriately, value added manufacturing should be the base of a healthy Singapore economy.

Why is this? Surely SPEC is part of the huge Sony corporation, part of Japan Inc, and where Japanese companies compete or locate they inevitably export their management practices and philosophies. This may be the case but what SPEC have done is look at the Japanese methodology, adapted it to the Singapore lifestyle and actually improved beyond the expectations of Sony Corporation.

Management as well as employees at SPEC feel they share the same fate and this appears to translate into a system of equality where differentials are kept to a minimum.

The argument could be raised that the workforce at SPEC was so young anyway that differentials would therefore not be apparent. This may be so, but what was apparent was (at the risk of being repetitive) respect for the individual.

With so much emphasis placed on individual performance it should be remembered that SPEC also place great importance on teamwork. Teams do exist on a cross functional basis but the real team is SPEC. The entire company pulls together, achieving a common objective, achieving results and building trust amongst all who participate.

Education and training are key factors to the success story at SPEC. Induction training, on the job training, product and process modification training are the norm. Training in quality practices was exceptional, with a high percentage of employees trained in cause and effect problem analysis techniques.

The lifetime employment system, common practice in Japan, exists also in Singapore due mainly to the labour shortage situation. With jobs available, in excess of labour supply the potential for job mobility or changeover is obviously increasing. However, the real situation at SPEC is that labour turnover is minimal, due to the obvious excellent working conditions and worker participation.

SPEC has realised that there is little point in investing in technology as a panacea for all its problems. Technology, without a corresponding increase in training the people of the organisation to accept technology, would not have worked at SPEC. Within SPEC there is a tremendous pool of creativity which has been harnessed and focused in a strategic direction. A greater understanding of technology has allowed SPEC and its employees the ability to make jobs better, faster, more efficient and easier to handle. The management at SPEC also understands the contribution its people make who in turn understand the objectives of the company.

SPEC is very much part of Singapore's strategic plan to maintain its strong position within the world economy. Singapore has taken a planned coordinated approach to attracting investment and is much more open to influence from Japanese technology than are its Western, or indeed some of its Asian neighbours. The infrastructure of Singapore: air, sea, communication and its geographic location, has ensured (at least to date) that Singapore has been able to maintain a competitive advantage. Coupled with a disciplined labour force and a propensity to concentrate on value added, high quality niche markets, this competitive advantage looks likely to continue in at least the foreseeable future.

Appendix 1: MISSION STATEMENT

Sony Precision Engineering Centre (S) Pte Ltd (SPEC) has a threefold function – to manufacture precision components; to provide specialist high-tech services in factory automation; and to act as a technological support base for Sony operations in the region.

But it has only one mission – to be a centre for excellence in all its activities.

This is underpinned by SPEC's total commitment to quality. In the world of precision engineering, only the highest quality is acceptable and SPEC achieves this by a combination of the most advanced automation and robotics with a skilled, well motivated workforce.

More than 80 per cent of SPEC operations are automated. The vital human element consists of specialist engineers and technicians working together as a cohesive and integrated team. The Singapore worker has been consistently judged the most productive and efficient in the world and is a major factor in the consistently high quality of SPEC's output.

Like its parent company, Sony Corporation of Japan, SPEC believes that creative thinking leads to innovative technology. By nurturing this philosophy, SPEC has not only become a focal point for technology transfer from Japan to ASEAN, but also pioneered technology and procedures on its own.

SPEC is a dynamic company, constantly growing in stature, expanding its operations. It shares with Sony the aspiration to develop new products, to refine existing products and ultimately to make a contribution to the way we live.

This case study was prepared by John McMorrow.

Case study questions

1 Evaluate the human resource management practices at SPEC in relation to the 'Sony Spirit'.

2 How important were the 'productivity month' and the '100% Club' at SPEC in providing motivation and rewards to employees?

Appendix A
Three American gurus

A small group of American quality experts or 'gurus' have, in the past, advised industry throughout the world on how it should manage quality. The approaches of Philip B Crosby, W Edwards Deming, and Joseph M Juran, their similarities and differences, are presented briefly here.

Philip B Crosby

Crosby's four absolutes of quality:

- Definition – conformance to requirements.
- System – prevention.
- Performance standard – zero defects.
- Measurement – price of non-conformance.

He offers management 14 steps to improvement:

1 Make it clear that management is committed to quality.
2 Form quality improvement teams with representatives from each department.
3 Determine where current and potential quality problems lie.
4 Evaluate the cost of quality and explain its use as a management tool.
5 Raise the quality awareness and personal conern of all employees.
6 Take actions to correct problems identified through previous steps.
7 Establish a committee for the zero defects programme.
8 Train supervisors to actively carry out their part of the quality improvement programme.
9 Hold a 'zero defects day' to let all employees realise that there has been a change.
10 Encourage individuals to establish improvement goals for themselves and their groups.
11 Encourage employees to communicate to management the obstacles they face in attaining their improvement goals.
12 Recognise and appreciate those who participate.
13 Establish quality councils to communicate on a regular basis.
14 Do it all over again to emphasise that the quality improvement programme never ends.

W Edwards Deming

Deming's 14 points for management are the following:

1 Create constancy of purpose towards improvement of product and service.
2 Adopt the new philosophy. We can no longer live with commonly accepted levels of delays, mistakes, defective workmanship.
3 Cease dependence on mass inspection. Require, instead, statistical evidence that quality is built-in.
4 End the practice of awarding business on the basis of price tags.
5 Find problems. It is management's job to work continually on the system.
6 Institute modern methods of training on the job.
7 Institute modern methods of supervision of production workers. The responsibility of foremen must be changed from numbers to quality.
8 Drive out fear, so that everyone may work effectively for the company.
9 Break down barriers between departments.
10 Eliminate numerical goals, posters, and slogans for the workforce asking for new levels of productivity without providing methods.
11 Eliminate work standards that prescribe numerical quotas.
12 Remove barriers that stand between the hourly worker and this right to pride of workmanship.
13 Institute a vigorous programme of education and training.
14 Create a structure in top management that will push the above 13 points everyday.

Joseph M Juran

Juran's 10 steps to quality improvement are the following:

1 Build awareness of the need and opportunity for improvement.
2 Set goals for improvement.
3 Organise to reach the goals (establish a quality council, identify problems, select projects, appoint teams, designate facilitators).
4 Provide training.
5 Carry out projects to solve problems.
6 Report progress.
7 Give recognition.
8 Communicate results.
9 Keep score.
10 Maintain momentum by making annual improvement part of the regular systems and processes of the company.

A comparison

One way to directly compare the various approaches of the three American gurus is in tabular form. Table A.1 shows the differences and similarities, classified under 12 different factors.

Table A.1 *The American quality gurus compared*

	Crosby	Deming	Juran
Definition of quality	Conformance to requirements	A predictable degree of uniformity and dependability at low cost and suited to the market	Fitness for use
Degree of senior-management responsibility	Responsible for quality	Responsible for 94% of quality problems	Less than 20% of quality problems are due to workers
Performance standard/motivation	Zero defects	Quality has many scales. Use statistics to measure performance in all areas. Critical of zero defects	Avoid campaigns to do perfect work
General approach	Prevention, not inspection	Reduce variability by continuous improvement. Cease mass inspection	General management approach to quality – especially 'human' elements
Structure	Fourteen steps to quality improvement	Fourteen points for management	Ten steps to quality improvement
Statistical process control (SPC)	Rejects statistically acceptable levels of quality	Statistical methods of quality control must be used	Recommends SPC but warns that it can lead to tool-driven approach
Improvement basis	A 'process', not a programme. Improvement goals	Continuous to reduce variation. Eliminate goals without methods	Project-by-project team approach. Set goals
Teamwork	Quality improvement teams. Quality councils	Employee participation in decision-making. Break down barriers between departments	Team and quality circle approach
Costs of quality	Cost of non-conformance. Quality is free	No optimum – continuous improvement	Quality is not free – there is an optimum

Table A.1 *(continued)*

	Crosby	Deming	Juran
Purchasing and goods received	State requirements. Supplier is extension of business. Most faults due to purchasers themselves	Inspection too late – allows defects to enter system through AQLs. Statistical evidence and control charts required	Problems are complex. Carry out formal surveys
Vendor rating	Yes *and* buyers. Quality audits useless	No – critical of most systems	Yes, but help supplier improve
Single sources of supply		Yes	No – can neglect to sharpen competitive edge.

Appendix B
Bibliography

General quality management and TQM

Bank, J, *The Essence of Total Quality Management*, Prentice Hall, Hemel Hempstead (UK), 1992.

Caplen, R H, *A Practical Approach to Total Quality Control* (5th edn), Business Books, London, 1988.

Crosby, P B, *Quality is Free*, McGraw-Hill, New York, 1979.

Crosby, P B, *Quality Without Tears*, McGraw-Hill, New York, 1984.

Dale, B G and Plunkett, J J (eds), *Managing Quality*, Philip Alan, Hemel Hempstead (UK), 1990.

Deming, W E, *Out of the Crisis*, MIT, Cambridge, Mass. (USA), 1982.

Deming, W E, *The New Economies*, MIT, Cambridge, Mass. (USA), 1983.

Edosomwam, J A, *Productivity and Quality Improvement*, IFS, Bedford (UK), 1988.

Feigenbaum, A V, *Total Quality Control* (3rd edn, revised), McGraw-Hill, New York, 1991.

Garvin, D A, *Managing Quality: the strategic competitive edge*, The Free Press (Macmillan), New York, 1988.

Hakes, C (ed), *Total Quality Improvement: the key to business improvement*, Chapman & Hall, London, 1991.

Hutchins, D, *In Pursuit of Quality*, Pitman, London, 1990.

Hutchins, D, *Achieve Total Quality*, Director Books, Cambrdige (UK), 1992.

Ishikawa, K (translated by D J Lu), *What is Total Quality Control?: the Japanese way*, Prentice-Hall, Englewood Cliffs, NJ (USA), 1985.

Macdonald, J and Piggot, J, *Global Quality: the new management culture*, Mercury Books, London, 1990.

Mann, N R, *The Keys to Excellence: the story of the Deming philosophy*, Prestwick Books, Los Angeles, CA (USA), 1985.

Murphy, J A, *Quality in Practice*, Gill and MacMillan, Dublin, 1986.

Popplewell, B and Wildsmith, A, *Becoming the Best*, Gower, Aldershot (UK), 1988.

Price, F, *Right Every Time*, Gower, Aldershot (UK), 1990.

Sarv Singh Soin, *Total Quality Control Essentials – key elements, methodologies and managing for success*, McGraw-Hill, New York, 1992.

Wille, E, *Quality: achieving excellence*, Century Business, London, 1992.

Zairi, M, *Total Quality Management for Engineers*, Woodhead, Cambridge (UK), 1991.

Leadership and commitment

Adair, J, *Not Bosses but Leaders: how to lead the successful way*, Talbot Adair Press, Guildford (UK), 1987.

Adair, J, *The Action-Centred Leader*, Industrial Society, London, 1988.
Adair, J, *Effective Leadership* (2nd edn), Pan Books, London, 1988.
Crosby, P B, *Running Things*, McGraw-Hill, New York, 1986.
Juran, J M, *Juran on Leadership for Quality: an executive handbook*, The Free Press (Macmillan), New York, 1989.
Townsend, P L and Gebhardt, J E, *Commit to Quality*, J Wiley Press, New York, 1986.
Townsend, P L and Gebhardt, J E, *Quality in Action – 93 lessons in leadership, participation and measurement*, J Wiley Press, New York, 1992.

Customers, suppliers and service

Albin, J M, *Quality Improvement in Employment and other Human Services – managing for quality through change*, Paul Brookes Pub. (USA), 1992.
Cook, S, *Customer Care – implementing total quality in today's service driven organization*, Kogan Page, London, 1992.
Groocock, J M, *The Chains of Quality*, John Wiley, Chichester (UK), 1986.
King Taylor, L, *Quality: total customer service* (a case study book), Century Business, London, 1992.
Lash, L M, *The Complete Guide to Customer Service*, J Wiley Press, New York, 1989.
Mastenbrock, W (ed), *Managing for Quality in the Service Sector*, Basil Blackwell, Oxford (UK), 1991.
Zeithaml, V A, Parasuraman, A and Berry, L L, *Delivering Quality Service: balancing customer perceptions and expectations*, The Free Press (Macmillan), New York, 1990.

Design, innovation and QFD

Adair, J, *The Challenge of Innovation*, Talbot Adair Press, Guildford (UK), 1990.
Fox J, *Quality Through Design*, MGLR, 1993.
Juran, J J, *Juran on Quality by Design*, Free Press, New York, 1992.
Marsh, S, Moran, J, Nakui, S and Hoffherr, G D, *Facilitating and Training in QFD*, ASQC, Milwaukee, WI (USA), 1991.
Zairi, M, *Management of Advanced Manufacturing Technology*, Sigma Press, Wilmslow (UK), 1992.

Quality planning, JIT, and POM

Ansari, A and Modarress, B, *Just-in-time Purchasing*, The Free Press (Macmillan), New York, 1990.
Bineno, J, *Implementing JIT*, IFS, Bedford (UK), 1991.
Harrison, A, *Just-in-Time Manufacturing in Perspective*, Prentice-Hall, Englewood Cliffs, NJ (USA), 1992.
Hutchins, D, *Just-in-Time*, Gower, Aldershot (UK), 1988.

Juran, J M (ed), *Quality Control Handbook*, McGraw-Hill, New York, 1988.

Juran, J M and Gryna, F M, *Quality Planning and Analysis* (2nd edn), McGraw-Hill, New York, 1980.

Muhlemann, A P, Oakland, J S and Lockyer, K G, *Production and Operations Management* (6th edn), Pitman, London, 1992.

Voss, C A (ed), *Just-in-Time Manufacture*, IFS Publications, Bedford, (UK), 1989.

Quality systems

Dale, B G and Oakland, J S, *Quality Improvement Through Standards*, (2nd edn), Stanley Thornes, Chelthenham (UK), 1994.

Hall, T J, *The Quality Manual – the application of BS5750 ISO 9001 EN 29001*, John Wiley, Chichester (UK), 1992.

Rothery, B, *ISO 9000*, Gower, Aldershot (UK), 1991.

Stebbing L, *Quality Assurance: the route to efficiency and competitiveness* (2nd edn), John Wiley, Chichester (UK), 1989.

The Baldrige and European Quality Award criteria

Brown M G, *Baldrige Award Winning Quality: how to interpret the Malcolm Baldrige Award criteria* (2nd edn), ASQC, Milwaukee, WI (USA), 1992.

EPQM (European Foundation for Quality Management), *The European Model for TQM – guide to self assessment*, 1995.

Hart, W L and Bogan, C E, *The Baldrige: what it is, how it's won, how to use it to improve quality in your company*, McGraw-Hill, New York, 1992.

Mills Steeples, M, *The Corporate Guide to the Malcolm Baldrige National Quality Award*, ASQC, Milwaukee, WI (USA), 1992.

NIST (US Dept of Commerce, National Institute of Standards and Technology), *Malcolm Baldrige National Quality Award Criteria*, 1995.

Quality costing, measurement and benchmarking

Bendell, Tony, *Benchmarking for Competitive Advantage*, Longman, 1993.

Camp, R C, *Benchmarking: the search for industry best practices that lead to superior performance*, ASQC Quality Press, Milwaukee, WI (USA), 1989.

Dale B G, and Plunkett, J J, *Quality Costing*, Chapman and Hall, London, 1991.

Dixon, J R, Nanni, A and Vollmann, T E, *The New Performance Challenge – measuring operations for world class competition*, Business One Irwin, Homewood (USA), 1990.

Hall, R W, Johnson, H Y and Turney, P B B, *Measuring Up – charting pathways to manufacturing excellence*, Business One Irwin, Homewood (USA), 1991.

Kaplan, R W (ed), *Measures for Manufacturing Excellence*, Harvard Business School Press, Boston, Mass. (USA), 1990.

Kinlaw, D C, *Continuous Improvement and Measurement for Total Quality – a team - based approach*, Pfieffer & Business One (USA), 1992.

Porter, L J and Rayne, P, 'Quality costing for TQM', *International Journal of Production Economics*, 27, pp. 69-81, 1992.

Spendolini, M J, *The Benchmarking Book*, ASQC, Milwaukee, WI (USA), 1992.

Talley, D J, *Total Quality Management: performance and cost measures*, ASQC, Milwaukee, WI (USA), 1991.

Zairi, M, *Competitive Benchmarking*, TQM Practitioner Series, Technical Communications (Publishing), Letchworth (UK), 1992.

Zairi, M, *TQM-Based Performance Measurement*, TQM Practitioner Series, Technical Communications (Publishing), Letchworth (UK), 1992.

Zairi, M, *Measuring Performance for Business Results*, Chapman and Hall, 1994.

Zairi, M and Leonard, P, *Practical Benchmarking – the complete guide*, Chapman and Hall, 1994.

Tools and techniques of TQM (including SPC)

Bhote, K R, *World Class Quality – using design of experiments to make it happen*, AMACOM, New York (USA), 1991.

Carlzon, J, *Moments of Truth*, Ballinger, Cambridge, Mass. (USA), 1987.

Caulcutt, R, *Data Analysis in the Chemical Industry, Vol. 1: basic techniques*, Ellis Horwood, Chichester (UK), 1989.

Caulcutt, R, *Statistics in Research and Development* (2nd edn), Chapman and Hall, London, 1991.

Joiner, B, *Fourth Generation Management*, McGraw-Hill, 1994.

Neave, H, *The Deming Dimension*, SPC Press, Knoxville (USA), 1990.

Oakland, J S and Followell, R F, *Statistical Process Control: a practical guide* (2nd edn), Butterworth-Heinemann, Oxford (UK), 1990.

Price, F, *Right First Time*, Gower, London, 1985.

Ryuka Fukuda, *CEDAC – a tool for continuous systematic improvement*, Productivity Press, Cambridge, Mass. (USA), 1990.

Wheeler, D, *Understanding Variation*, SPC Press, 1993.

Shingo and Taguchi methods

Bendell, T, Wilson, G and Millar, R M G, *Taguchi Methodology with Total Quality*, IFS, Bedford (UK), 1990.

Lagothetis, N, *Managing for Total Quality – from Deming to Taguchi and SPC*, Prentice-Hall, Englewood Cliffs, NJ (USA), 1990.

Ranjit, Roy, *A Primer on the Taguchi Method*, Van Nostrand Reinhold, New York, 1990.

Shingo, S, *Zero Quality Control: source inspection and the Poka-yoke system*, Productivity Press, Stamford, Conn. (USA), 1986.

TQM through people and teamwork

Adair, J, *Effective Teambuilding* (2nd edn), Pan Books, London, 1987.

Aubrey, C A and Felkins, P K, *Teamwork: involving people in quality and productivity improvement*, ASQC, Milwaukee, WI (USA), 1988.

Belbin, R M, *Management Teams: why they succeed or fail*, Butterworth-Heinemann, Oxford (UK), 1981.

Blanchard, K and Hersey, P, *Management of Organizational Behaviour: utilizing human resources* (4th edn), Prentice-Hall, Englewood Cliffs, NJ (USA), 1982.

Briggs Myers, I, *Introduction to Type: a description of the theory and applications of the Myers Briggs Type Indicator*, Consulting Psychologists Press, Palo Alto (USA), 1987.

Choppin, J, *Quality Through People: a blueprint for proactive total quality management*, IFS, Kempston (UK), 1991.

Collard, R, *Total Quality: success through people*, Institute of Personnel Management, Wimbledon (UK), 1989.

Dale, B G and Cooper, C, *Total Quality and Human Resources – an executive guide*, Blackwell, Oxford (UK), 1992.

Hutchins, D, *The Quality Circle Handbook*, Gower, Aldershot (UK), 1985.

Kormanski, C, 'A situational leadership approach to groups using the Tuckman Model of Group Development', *The 1985 Annual: developing human resources*, University Associates, San Diego (USA), 1985.

Kormanski, C and Mozenter, A, 'A new model of team building: a technology for today and tomorrow', *The 1987 Annual: developing human resources*, University Associates, San Diego (USA), 1987.

Krebs Hirsh, S, *MBTI Team Building Program, Team Member's Guide*, Consulting Psychologists Press, Palo Alto, CA (USA), 1992.

Krebs Hirsh, S and Kummerow, J M, *Introduction to Type in Organizational Settings*, Consulting Psychologists Press, Palo Alto, CA (USA), 1987.

McCaulley, M H, 'How individual differences affect health care teams', *Health Team News*, 1(8), pp. 1-4, 1975.

Masaaki, I, *Kaizen: the key to Japanese competitive success*, McGraw-Hill, New York, 1986.

Robson, M, *Quality Circles: a practical guide* (2nd edn), Gower, Aldershot (UK), 1989.

Scholtes, P R, *The Team Handbook*, Joiner Associates, Madison, NY (USA), 1990.

Shetty, Y K and Buehler, V M (eds), *Productivity and Quality Through People* (case studies), Quorum Books, London, 1985.

Tannenbaum, R and Schmidt, W H, 'How to choose a leadership pattern', *Harvard Business Review*, May-June, 1973.

Tuckman, B W and Jensen, M A, 'Stages of small group development revisited', *Group and Organizational Studies*, 2 (4), pp. 419-427, 1977.

Wellins, R S, Byham, W C and Wilson, J M, *Empowered Teams*, Jossey Bass, Oxford (UK), 1991.

Whitley, R, *The Customer Driven Company*, Business Books, London, 1991.

Cross-functional process improvement

Dimaxcescu, D, *The Seamless Enterprise – making cross-functional management work*, Harper Business, New York, 1992.

Francis, D, *Unblocking the Organisational Communication*, Gower, Aldershot (UK), 1990.

Hammer, M and Champy, J, *Re-engineering the Corporation*, Nicholas Brearley, 1993.

Harrington, H J, *Business Process Improvement*, McGraw-Hill, New York, 1991.

Rummler, G A and Brache, A P, *Improving Performance: how to manage the white space on the organisation chart*, Jossey-Bass Publishing, San Francisco, CA (USA), 1990.

Senge, P M, *The Fifth Discipline*, Century Business, 1990.

Implementing TQM

Atkinson, P E, *Creating Culture Change: the key to successful total quality management*, IFS, Bedford (UK), 1990.

Ciampa, D, *Total Quality – a user's guide for implementation*, Addison-Wesley, Reading, Mass. (USA), 1992.

Crosby, P B, *The Eternally Successful Organization*, McGraw-Hill, New York, 1988.

Cullen, J and Hollingham, J, *Implementing Total Quality*, IFS (Publications), London, 1987.

Fox, R, *Six Steps to Total Quality Management*, McGraw-Hill, NSW (Australia), 1991.

Gitlow, H S and Gitlow, S J, *The Deming Guide to Quality and Competitive Position*, Prentice-Hall, New Jersey (USA), 1987.

Hardaker, M and Ward, B K, 'Getting things done – how to make a team work', *Harvard Business Review*, pp. 112-119, Nov/Dec 1987.

Hiam, A, *Closing the Quality Gap – lessons from America's leading companies*, Prentice-Hall, Englewood Cliffs, NJ (USA), 1992.

Morgan, C and Murgatroyd, S, *Total Quality Management in the Public Sector*, Open University Press, 1994.

Munro-Faure, L and Munro-Faure, M, *Implementing Total Quality Management* , Pitman, London, 1992.

Saylor, J H, *Total Quality Management Field Manual*, McGraw-Hill, New York, 1992.

Scherkenbach, W W, *The Deming Route to Quality and Productivity: road maps and road blocks*, Mercury Press/Fairchild Publications, Rockville, Md (USA), 1986.

Scherkenbach, W W, *Deming's Road to Continual Improvement*, SPC Press, Knoxville (USA), 1991.

Schuler, R S and Harris, D L, *Managing Quality – the primer for middle managers*, Addison-Wesley, Reading, Mass. (USA), 1992.

Spenley, P, *World Class Performance through Total Quality*, Chapman and Hall, London, 1992.

Tennor, A R and De Toro, I J, *Total Quality Management – three steps to continuous improvement*, Addison-Wesley, Reading, Mass. (USA), 1992.

Tunks, R, *Fast Track to Quality*, McGraw-Hill, New York, 1992.

Index